International Conference on Material Science and Engineering Technology (ICMSET 2021)

IOP Conference Series: Materials Science and Engineering
Volume 1150

Jeju Island, South Korea
12 - 14 March 2021

ISBN: 978-1-7138-3420-5
ISSN: 1757-8981

Printed from e-media with permission by:

Curran Associates, Inc.
57 Morehouse Lane
Red Hook, NY 12571

Some format issues inherent in the e-media version may also appear in this print version.

Institute of Physics
Dirac House, Temple Back
Bristol BS1 6BE UK

Phone: 44 1 17 929 7481
Fax: 44 1 17 920 0979

techtracking@iop.org

Additional copies of this publication are available from:

Curran Associates, Inc.
57 Morehouse Lane
Red Hook, NY 12571 USA
Phone: 845-758-0400
Fax: 845-758-2633
Email: curran@proceedings.com
Web: www.proceedings.com

TABLE OF CONTENTS

IOP Conf. Series: Materials Science and Engineering 1150 (2021) 011001 doi:10.1088/1757-899X/1150/1/011001

Preface

The ICMSET 2021 is the 5th International Conference on Material Science and Engineering Technology (5th ICMSET 2021) will take place in Jeju Island, Korea during March 12-14, 2021. But currently, the entire world is struggling against the virulent pandemic COVID-19, so that the ICMSET 2021 conference changed the presentation method from Oral presentation to Remote Video Presentation to help with Global Quarantine Activities.

The Virtual of ICMSET 2021 was held on Jeju Island, Korea at Video Conference Meeting Room on March 12-14, 2021. The ICMSET Committee members attended conference at on-site and also authors attended the conference as Video Meeting. The conference program covered Invited Keynote Speech, Oral, and Poster Presentations, they presented their presentation 15 minutes including Q&A time.

The committee of ICMSET 2021 expresses their sincere thanks to all authors for their high-quality research papers and careful presentations. All reviewers are also thanked for their careful comments and advices. Thanks, are finally given to IOP Publication as well for producing this volume.

The Organizing Committee of ICMSET 2021

Committee Chair

Prof. Dongkeon Kim

Dong-A University

IOP Conf. Series: Materials Science and Engineering 1150 (2021) 011001 doi:10.1088/1757-899X/1150/1/011001

Conference Photograph

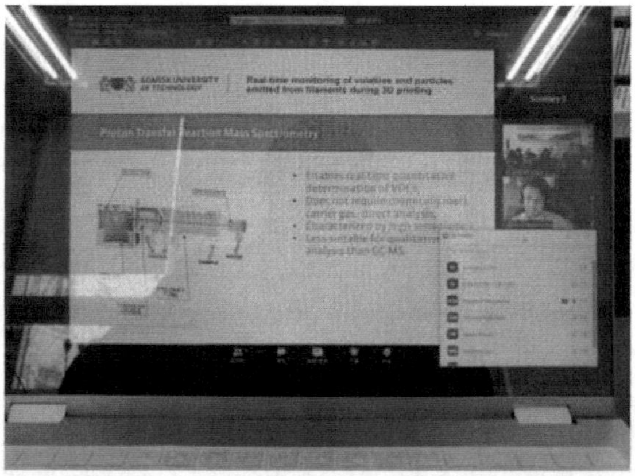

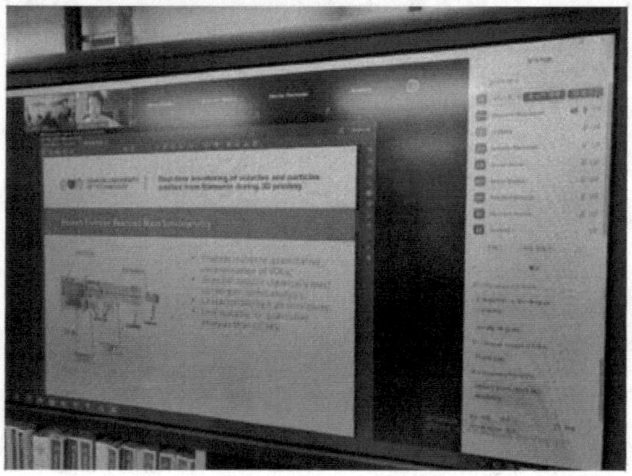

IOP Conf. Series: Materials Science and Engineering 1150 (2021) 011001 doi:10.1088/1757-899X/1150/1/011001

Sponsors

Incheon Disaster Prevention Research Center (IDPRC)	IDPRC 인천대학교 방재연구센터 Incheon Disaster Prevention Center

Recently, various efforts to prevent and prepare are vitally needed for prevention of Disasters and calamities. So we understand the necessity for technology of disaster and we built up the Incheon Disaster Prevention Research Center (IDPRC) in Incheon National University (INU) in 1997.

Accordingly, Incheon Disaster Prevention Research Center (IDPRC) in Incheon National University has progressed of research on the prevention of disasters and calamities through the various seminars, conference and lectures. This research could be conducted cooperation with Incheon National University (INU) in various fields' structure, soil, hydraulic and environment.

Incheon Disaster Prevention Research Center (IDPRC) will try to be a leader in the disaster of industry through the various research activities and global conference.

University Core Research Center for Disaster-free and Safe Ocean City Construction (D-SOC) at Dong-A University

Recently, various efforts to prevent and prepare are vitally needed for prevention of Disasters and calamities. So we understand the necessity for technology of disaster and we built up the University Core Research Center for Disaster-free and Safe Ocean City Construction (D-SOC) was built in 1965 in Dong-A University in order to contribute the development of construction industry.

Accordingly, University Core Research Center for Disaster-free and Safe Ocean City Construction (D-SOC) in Dong-A University has progressed of research on the construction technology and planning through the various seminars, conference and lectures. These researches could be conducted cooperation with Dong-A University in various fields structure, soil, hydraulic, environment.

University Core Research Center for Disaster-free and Safe Ocean City Construction (D-SOC) will try to be a leader in the construction of industry through the various research activities, technical exchanges and global conference.

IOP Conf. Series: Materials Science and Engineering 1150 (2021) 011001 doi:10.1088/1757-899X/1150/1/011001

Lists of Committees

Committee name	Organizer name	Organization
Honorary Chair	Jongwan Hu	Incheon National University
Conference General Chairs	Mosbeh Kaloop	Incheon National University
Honorary Chair	Dongkeon Kim	Dong-A University
Technical Committees	J. W. Hu	Incheon National University

IOP Conf. Series: Materials Science and Engineering 1150 (2021) 011002 doi:10.1088/1757-899X/1150/1/011002

Peer review declaration

All papers published in this volume of *IOP Conference Series: Materials Science and Engineering* have been peer reviewed through processes administered by the Editors. Reviews were conducted by expert referees to the professional and scientific standards expected of a proceedings journal published by IOP Publishing.

- **Type of peer review:**
  All papers revise by three of committee members of the conference. The papers are divided based on the specialist of the committee members; and in the case of special topics, other reviewers invite for reviewing the selected manuscripts.

  Also, reviewers have considered the Technical Criteria (Scientific merit: notably scientific rigour, accuracy and correctness. Clarity of expression; communication of ideas; readability and discussion of concepts. Sufficient discussion of the context of the work, and suitable referencing) and Quality Criteria (Originality: the work relevant and novel, Motivation: Does the problem considered have a sound motivation? All papers should clearly demonstrate the scientific interest of the results. Repetition: Have significant parts of the manuscript already been published? Length: Is the content of the work of sufficient scientific interest to justify its length?)

- **Conference submission management system:**
  Our conference has Secretary for management on conference papers. So, authors send their papers to Conference E-mail, and our secretary has managed it. The secretary sends the papers to the conference chairs to evaluate the suitability of submitted papers to the conference scopes.

- **Number of submissions received:**
  Especially in this time, only few of papers were submitted cause of COVID-19. The total number of submission papers were 110 papers.

- **Number of submissions sent for review:**
  The submission papers were revised from the management conference system and committee organizations and the papers that sent for review were 68 papers.

- **Number of submissions accepted:**
  The papers that accepted in the conference for publishing after review processing are 25 papers.

- **Acceptance Rate (Number of Submissions Accepted / Number of Submissions Received X 100):** 22 %

- **Average number of reviews per paper:**
  Different cycles review processing implemented for improving the papers quality. In average, three times review on per paper were processed.

- **Total number of reviewers involved:**

Three of reviewers are took part in the review process. When we receive positive reports from two at least we can accept the paper after evaluating the reviewer report and authors responses.

- **Any additional info on review process:**
  The processing of review work of this conference was carried out as follows: (1) the secretary received the papers by emails and processed the initial checking of the papers, he revised the format, language and roughly check of papers contents; (2) the general chairs received the processed papers and they revised the papers scopes, Technical Criteria (Scientific merit: notably scientific rigour, accuracy and correctness. Clarity of expression; communication of ideas; readability and discussion of concepts. Sufficient discussion of the context of the work, and suitable referencing) and Quality Criteria (Originality: the work relevant and novel, Motivation: Does the problem considered have a sound motivation? All papers should clearly demonstrate the scientific interest of the results. Repetition: Have significant parts of the manuscript already been published? Length: Is the content of the work of sufficient scientific interest to justify its length?)
  If these papers relate to the conference scopes it can be passed to the next processing, else the papers can be rejected. (3) all accepted papers divided into categories based on the papers contents and sent to three committee members who have a good experience in the paper contents. (4) after receiving the review reports from the committee members, the review reports sent to authors for sending their responses on the listed comments. (5) the reviewing cycling process divided into accept, minor and major revisions; and based on the review comments and suggestion, the papers can be processing up to three times to improve the manuscript contents. (6) finally, the authors should revise/argue the comments and send their response through email; and the processing cycling until organizer accept the paper according to the final review reports.

- **Contact person for queries:**
  : Jack Son (E-mail: icmset2021@icmset-conference.org)

IOP Conf. Series: Materials Science and Engineering 1150 (2021) 012001 doi:10.1088/1757-899X/1150/1/012001

Real-time monitoring of volatiles and particles emitted from thermoplastic filaments during 3D printing

Wojciech Wojnowski[1,a*], Renata Marcinkowska[1,b], and Bożena Zabiegała[1,c]

[1]Department of Analytical Chemistry, Faculty of Chemistry, Gdańsk University of Technology, 11/12 Narutowicza Street, 80-233 Gdańsk, Poland

E-mail: wojciech.wojnowski@pg.edu.pl

Abstract: The proliferation of consumer-grade three-dimensional (3D) printers using fused deposition, also known as fused filament fabrication, has given rise to concerns over the exposure of users to potentially harmful substances. Thermoplastic filaments made of different polymers are extruded through a heated printer nozzle and deposited layer by layer on a build platform to form the printed object. This process leads to the emission of both particles and gases, in particular volatile organic compounds (VOCs). While the emission of airborne particulate matter during 3D-printing has been relatively well researched [1], this is not the case with the monitoring of the associated emission of VOCs, and so we are far from understanding the emission profile of 3D-printers as a whole. The main reason for that is the difficulty with sampling the emitted volatiles with sufficient time resolution to investigate e.g. the effect of rapid heating of the nozzle, and so the available data mostly describes either the total emission of particular VOCs during printing, or the time-resolved emission of total VOCs, without the determination of particular compounds [2,3]. In order to bridge this gap, we have previously demonstrated that proton transfer reaction mass spectrometry (PTR-MS) can be used to monitor the emission of volatiles during 3D-printing in real-time, both qualitatively and quantitatively [4]. Here, we propose to combine this approach with the concurrent real-time measurement of the corresponding emission of particles to obtain a more detailed understanding of the profile of pollutants generated during 3D-printing and the factors which determine this process.

1. Materials and methods

1.1 3D printer.

A commercial, consumer-grade FDM 3D printer (Prusa i3 MK2S, Prusa Research a.s., Prague, Czech Republic) was used in the study. The printer was equipped with a 0.4 mm nozzle and housed in an enclosure with a total volume of 130 dm^3. The air within the enclosure was mixed using three 120 mm fans. Four different filaments were used for the printing, namely polyethylene terephthalate glycol (PETG), polylactic acid (PLA) Ecoline (standard) (Print-Me, Gorzów Wielkopolski, Poland), PLA Premium, and acrylonitrile butadiene styrene (ABS) (Nebula Filaments, Stare Bystre, Poland). All four filaments were of the "natural" type, i.e. without added pigments or colourants. The test print was a 10 mm cube with rounded corners, with 3 top, bottom and wall layers, and 30% infill. The size of the cube was adjusted to obtain a printing time of approx. 5 min. The print speed was set to 40 mm/s, travel speed to 100 mm/s, and layer thickness to 0.2 mm. The nozzle and build plate temperature was set to 235°C in the case of PETG and ABS, and 220°C in the case of PLA, while the build platform temperature was set to 60°C. The model was sliced using Cura v.3.2.1 (Ultimaker, Utrecht, Netherlands). The volatiles emitted during printing were sampled into the PTR-MS via a capillary enclosed in a transfer line heated to 70°C at 100 cm^3 min^{-1}. In order to condition the capillaries prior to each measurement they were flushed with tetrahydrofuran in the case of PLA and PETG, and with acetone in the case of ABS, and incubated at 90°C for at least an hour. The particulates were sampled into the CPC (condensation particle counter) at 600 cm^3 min^{-1} via a short segment of dedicated conductive tubing. A schematic representation of the experimental setup is shown in Fig.1. The prints with each filament type and the corresponding measurements were carried out in quadruplicate in an air-conditioned laboratory with the ambient temperature kept at 20°C. The enclosure was opened and purged with clean air after each measurement until a baseline concentration of VOCs and particulates was reached, and the printer nozzle and build platform reached the starting temperature. A background measurement during which the 3D printer was operated without a filament was carried out before each round of four prints and subtracted from the results of measurements for each filament.

1.2 Proton transfer reaction mass spectrometry.
The changes in the composition of the atmosphere within the enclosure during 3D-printing with various polymer filaments were monitored using the PTR TOF 1000 Ultra (IoniconGmbH, Innsbruck, Austria) PTR-MS coupled with a time-of-flight detector. The voltage in the drift chamber was set to 610 V which translated to the E/N value of 120 Td (E/N is the field strength of the PTR drift tube, 1 Td = 10^{-17} V cm^2 $molecule^{-1}$). The concentration of the monitored volatiles was determined based on the kinetics of the proton transfer reaction with the hydronium ion and the k reaction rate of 2×10^{-9} cm^3 s^{-1}, corrected for the isotopic abundances [5]. The mass spectra were recorded throughout each measurement in 1 s increments using IoniTOF v. 3.0.76 and processed using PTR-MS Viewer v. 3.3.9.1 (IoniconGmbH, Innsbruck, Austria). The principle of operation of proton transfer mass spectrometry is discussed in more detail by Ellis and Mayhew [6].

1.3 Particle count.
The emission of particulates larger than 2 nm during 3D-printing was monitored using the Versatile Water-based Condensation Particle Counter model 3789 (TSI, Shoreview, Minnesota, USA) with critical orifice pressure of 56 kPa, and aerosol nozzle pressure of 3.9 kPa. The number concentration of particulates was averaged every second throughout the experiment and recorded using Aerosol Instrument Manager v. 2.3.0 (TSI, Shoreview, Minnesota, USA).

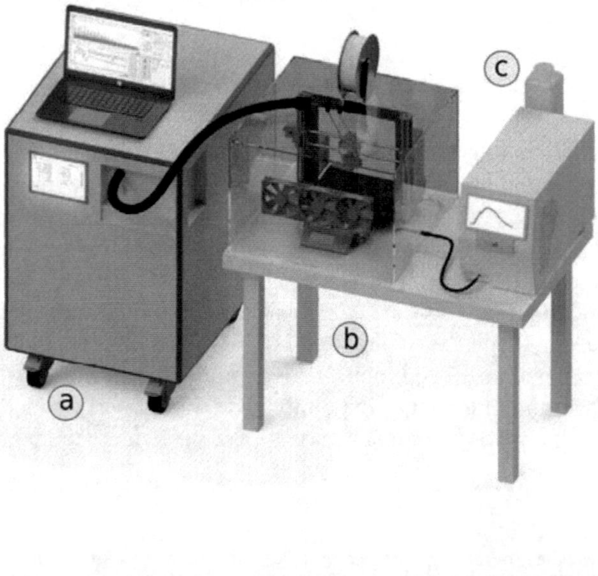

Figure 1. (previous page) The experimental setup consisting of a PTR-TOF-MS device with a heated transfer line (a), a chamber fitted with fans to facilitate air mixing within, enclosing an FDM 3D-printer (b), and a CPC (c).

2. Results and discussion

The main VOCs monitored in experiments with each of the four filaments were selected based on a 5 ppmv threshold. These were acetaldehyde (a product of secondary reactions of the most reactive VOCs, e.g. alkenes and dienes [7]) and the main products of thermal degradation of each polymer, i.e. lactide ($C_6H_8O_4H^+$, *m/z*=145 Th) and styrene ($C_8H_8H^+$, *m/z*=105 Th) in the case of PLA and ABS, respectively [8–10], and HC_2O^+ (*m/z*=42 Th), the product of dehydration of glyoxylic acid hydrate in the drift tube of PTR-MS, itself product of PETG degradation [11,12]. The changes in the relative concentration (after baseline correction) of the monitored VOCs and the corresponding concentration of particulates throughout printing are shown in Fig.2. It can be seen that the emission of main volatiles is not closely correlated with the overall emission of aerosol >2 nm. The concentration of VOCs increases steadily as the nozzle reaches the operating temperature and further during the printing until the printed object is completed due to the thermal degradation of the polymer matrices of the filaments. In the case of particulates, especially when printing using PETG and ABS, there is a sharp concentration increase as the nozzle temperature reaches the threshold value and a relatively large section of the filament within it is melted without the extruder motor yet activated, that is without the filament being continuously passed through the nozzle. When the printing starts and only a short segment of the filament is being melted in the nozzle at any given time the number concentration within the enclosure decreases, only to rise steadily throughout the printing as the particulates accumulate within it. This confirms the earlier reports that the heating process is the main contributor to the emission of particles, as opposed to the printing itself [13]. The higher and less uniform emission of particles when printing with "premium" PLA compared to printing with the less expensive product, and the corresponding lower emission rate of the main VOC, i.e. lactide (see Fig.2. and Tab.1.), is likely because the "premium" filament contains a larger proportion of additives such as plasticizers and flame retardants intended to improve the quality of the print [14].

Table 1. Average emission of particulates and selected volatiles during 3D-printing with four different polymer filaments per gram of filament used.

Filament	Acetaldehyde emission [cm^3 g^{-1}]	Main VOC* emission [cm^3 g^{-1}]	Particulates emission [10^{10} # g^{-1}]
PETG	60 ± 20	36 ± 11	19.8 ± 3.6
PLA Ecoline	32 ± 10	14.4 ± 5.6	5.2 ± 1.4
PLA Premium	24.3 ± 8.2	8.4 ± 3.7	13.1 ± 1.9
ABS	40 ± 14	59 ± 12	17.45 ± 0.60

* main VOCs: HC_2O^+, $C_4H_8O_4H^+$, and $C_8H_8H^+$ for PETG, PLA, and ABS, respectively.

Since the particles generated during 3D printing are a mixture of solids and liquids, the latter possibly formed by semi-volatile organic compounds (SVOCs) [15], and also of volatiles, it is necessary to use a variety of analytical techniques to comprehensively assess the potential exposure associated with 3D-printing. While the differences in overall emissions of both particulates and volatiles between different filament materials are significant and correlated, the same cannot be said about the relationship between the emission of particular VOCs and the corresponding total emission of particles. In order to investigate the dynamic changes in the emission rates during 3D-printing in more detail, it might be necessary to include another dimension in the analysis, i.e. the size of the measured particles, e.g. by combining CPC with an electrostatic classifier. This could reveal the relationship between the temporally-resolved composition of the aerosol within the printer enclosure with the corresponding concentration of VOCs determined in real-time using the PTR-MS. The proposed approach is a step towards obtaining a comprehensive emission profile of 3D-printers which could help to better understand the risk of exposure associated with using this technology and the relationship between the composition of the thermoplastic filaments and the emission of both volatiles and particulates.

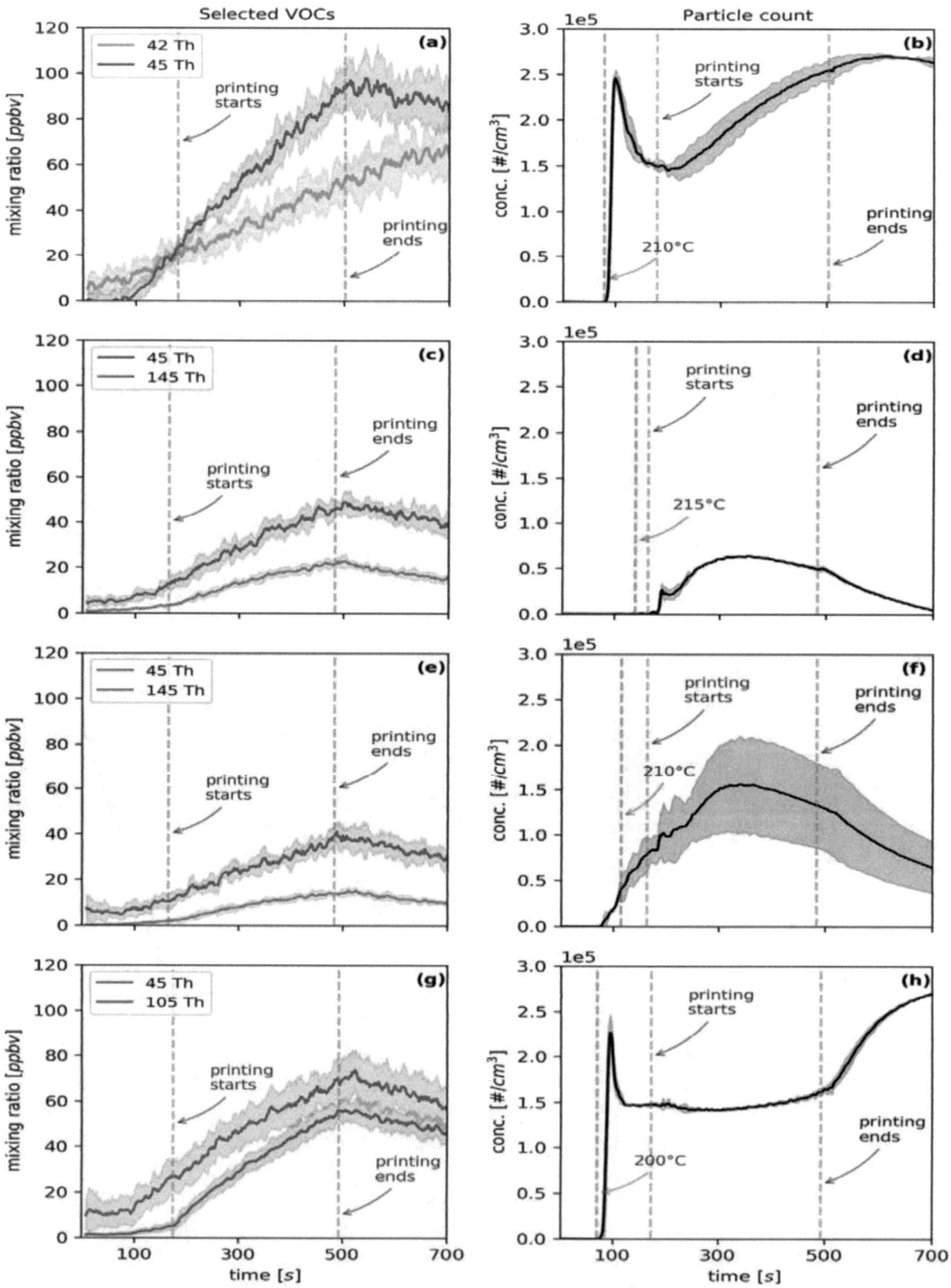

IOP Conf. Series: Materials Science and Engineering 1150 (2021) 012001 doi:10.1088/1757-899X/1150/1/012001

Figure 2. (previous page) Changes of the relative concentration (mixing ratio) ± SD of selected VOCs (protonated pseudo-molecular ion, MH^{+}) of PET-G (a), PLA Ecoline (c), PLA Premium (e), and ABS, and the corresponding changes of the concentration of particulates (b, d, f, h, respectively) within an enclosure during FDM 3D-printing.

Acknowledgements

This research was funded by the National Science Centre (Poland), grant number 2018/31/N/ST4/00809.

References

[1] P. Byrley, B.J. George, W.K. Boyes, K. Rogers, Particle emissions from fused deposition modeling 3D printers: Evaluation and meta-analysis, Sci. Total Environ. 655 (2019) 395–407.
[2] P. Steinle, Characterization of emissions from a desktop 3D printer and indoor air measurements in office settings, J. Occup. Environ. Hyg. 13 (2016) 121–132.
[3] E.L. Floyd, J. Wang, J.L. Regens, Fume emissions from a low-cost 3-D printer with various filaments, J. Occup. Environ. Hyg. 14 (2017) 523–533.
[4] W. Wojnowski, K. Kalinowska, J. Gębicki, B. Zabiegała, Monitoring the BTEX Volatiles during 3D Printing with Acrylonitrile Butadiene Styrene (ABS) Using Electronic Nose and Proton Transfer Reaction Mass Spectrometry, Sensors. 20 (2020) 5531.
[5] R.S. Blake, P.S. Monks, A.M. Ellis, Proton-transfer reaction mass spectrometry, Chem. Rev. 109 (2009) 861–896. https://doi.org/10.1021/cr800364q.
[6] A.M. Ellis, C.A. Mayhew, Proton Transfer Reaction Mass Spectrometry: principles and applications, John Wiley & Sons, Chichester, 2014.
[7] M. Possanzini, V. Di Palo, A. Cecinato, Sources and photodecomposition of formaldehyde and acetaldehyde in Rome ambient air, Atmos. Environ. 36 (2002) 3195–3201.
[8] F.-D. Kopinke, M. Remmler, K. Mackenzie, M. Möder, O. Wachsen, Thermal decomposition of biodegradable polyesters—II. Poly(lactic acid), Polym. Degrad. Stab. 53 (1996) 329–342.
[9] A.Y. Davis, Q. Zhang, J.P.S. Wong, R.J. Weber, M.S. Black, Characterization of volatile organic compound emissions from consumer level material extrusion 3D printers, Build. Environ. 160 (2019) 106209.
[10] S. Wojtyła, P. Klama, K. Śpiewak, T. Baran, 3D printer as a potential source of indoor air pollution, Int. J. Environ. Sci. Technol. 17 (2020) 207–218.
[11] B. Belloncle, F. Burel, H. Oulyadi, C. Bunel, Study of the in vitro degradation of poly(ethyl glyoxylate), Polym. Degrad. Stab. 93 (2008) 1151–1157.
[12] E. Boscaini, T. Mikoviny, A. Wisthaler, E. Von Hartungen, T.D. Märk, Characterization of wine with PTR-MS, Int. J. Mass Spectrom. 239 (2004) 215–219.
[13] Y. Deng, S.J. Cao, A. Chen, Y. Guo, The impact of manufacturing parameters on submicron particle emissions from a desktop 3D printer in the perspective of emission reduction, Build. Environ. 104 (2016) 311–319.
[14] J. Gu, M. Wensing, E. Uhde, T. Salthammer, Characterization of particulate and gaseous pollutants emitted during operation of a desktop 3D printer, Environ. Int. 123 (2019) 476–485.
[15] M.E. Vance, V. Pegues, S. Van Montfrans, W. Leng, L.C. Marr, Aerosol Emissions from Fuse-Deposition Modeling 3D Printers in a Chamber and in Real Indoor Environments, Environ. Sci. Technol. 51 (2017) 9516–9523.

Development of a cost-effectiveness model for a Mixed Municipal Waste Incinerators for Waste-to-energy (WtE) in small town

L Gotal Dmitrović[1*] and D Čerepinko[2]

[1] Associate Professor, University North, University Center Varazdin, Križaničeva 31b, Varaždin, Croatia, EU, lgotaldmitrovic@unin.hr
[2] Assistant Professor, University North, University Center Varazdin, Križaničeva 31b, Varaždin, Croatia, EU

E-mail: darijo.cerepinko@unin.hr

Abstract Waste-to-energy (WtE) or energy-from-waste (EfW) is the process of generating electricity and/or heat from the treatment of waste. Most WtE processes generate electricity and/or heat directly through waste combustion. According to the waste management hierarchy this method is better than disposal but worse than recycling. Many large cities use this method because it reduces the waste problems. In this paper, a cost-effectiveness model for Mixed Municipal Waste Incinerators is developed for a town, dealing with waste management methods problems for decades. Model development begins with the development of a cause-and-effect model, better known as the Ishikawa diagram. The paper develops mathematical dependences for: the dependence of specific investments on the capacity; the dependence of investment price on the capacity; the dependence of investment price on the capacity; the dependence of operating and maintenance costs on the capacity and the dependence of total plant efficiency on the capacity.
Keywords: Ishikawa diagram, costs, efficiency, internal rate of return

1. Introduction

With social and technological development, energy consumption is growing. A proven method of improving energy systems is decentralization of electricity production, which reduces energy losses associated with transmission and distribution. Most cities have significant potential to increase renewable energy production, using solar energy, biomass and converting waste into energy. The city of Freiburg has used the city's anaerobic power plant for the past decade, which annually converts 36,000 t of household organic waste into 4 million m3 of biogas and 15,000 t of fertilizer. Biogas is cleaned and then burned in cogeneration consumption, generating 7 GWh of electricity per year - which is a third of the city's energy produced from renewable sources - and transfers the heat to the heating network. One of the oldest central heating systems is the one in Copenhagen, which meets 97% of the city's heating needs. The heating network was initially based on coal, but the development of a central heating system has allowed the city to develop a much broader mix of heat sources - from natural gas, oil and coal to waste incineration, geothermal energy, biomass and straw. Cities like Stockholm and Mannheim provide heat using biofuels. In the future, the supply of electricity and heat in cities should be secured as much as possible from renewable energy sources. In this sense, the decisive factor for their application is the successful restructuring of the energy supply system.

IOP Conf. Series: Materials Science and Engineering 1150 (2021) 012002 doi:10.1088/1757-899X/1150/1/012002

2. Costs for a Mixed Municipal Waste Incinerators for Waste-to-energy (WtE)

The main causes of cost-effectiveness, derived from available secondary data on current European cases, are shown in the Ishikawa diagram (Figure 1). Ishikawa diagrams are used for Cause and Effect Analysis. This diagram-based technique, which combines Brainstorming with a type of Mind Map shows the specific causes of an effect. Causes are usually grouped into major categories to identify sources of variation [1].

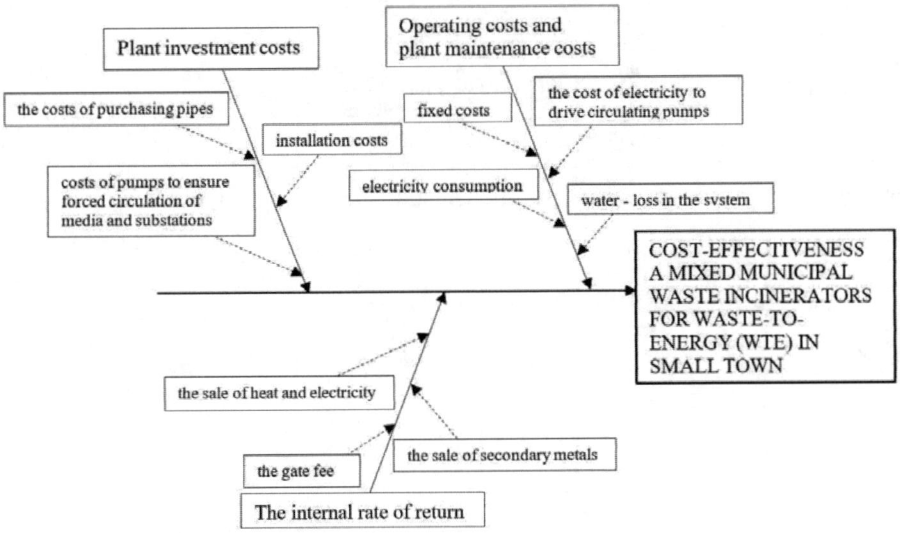

Figure. 1 Ishikawa diagram

2.1. Plant investment costs

Investment costs in the heating system distribution network consist of the costs of purchasing pipes and installation costs. In addition to the costs of the network itself, there are also costs of pumps to ensure forced circulation of media and substations that serve to transform thermal energy of higher parameters, distributed through pipelines, into thermal energy of the temperature level needed to meet the needs of the end user. The investment costs of the distribution network can be shown as a cost per meter of network or as a cost per heat capacity of the network. [2] Table 1 shows investment costs per ton of waste.

Table 1. Overview of plant investment costs

Capacity (t/y)	Year of construction	Investment price (€)	Specific investment (€/t)	References
40.000	2006.	24.248.000,00	606,20	[3]
100.000	2010.	56.000.000,00	560,00	[4]
200.000	2001.	107.000.000,00	535,00	[5] [6]
300.000	2006.	95.437.000,00	318,12	[3]
460.000	2007.	302.000.000,00	656,52	[7]
800.000	2004.	401.570.000,00	501,96	[7]

Since the years of construction differ, and thus the value of money, inflation is taken into account, and 2020 is taken as the reference year. Increased amounts are shown for the inflation rate in the country where the plant is located (Table 2).

Table 2. Overview of plant investment costs - calculated for 2020

Capacity (t/y)	Year of construction	Investment price (€)	Specific investment (€/t)
40.000	2006.	24.755.510,64	618,89
100.000	2010.	62.309.333,33	623,09
200.000	2001.	144.949.333,33	724,75
300.000	2006.	115.411.963,40	384,71
460.000	2007.	354.991.177,78	771,72
800.000	2004.	508.655.333,33	635,82

The dependence of specific investment on the capacity is shown in Figure 2.

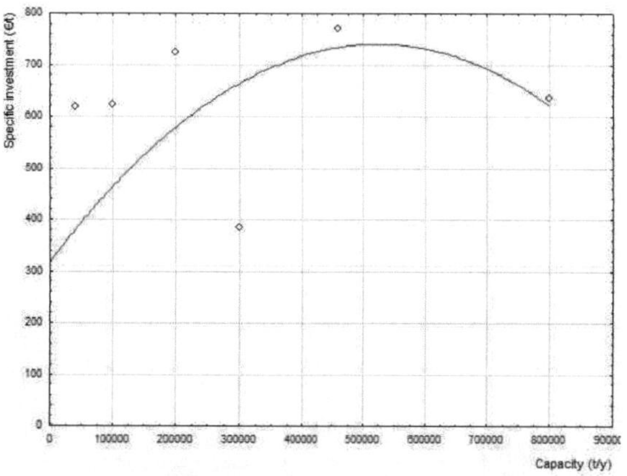

Figure. 2 The dependence of specific investments on the capacity

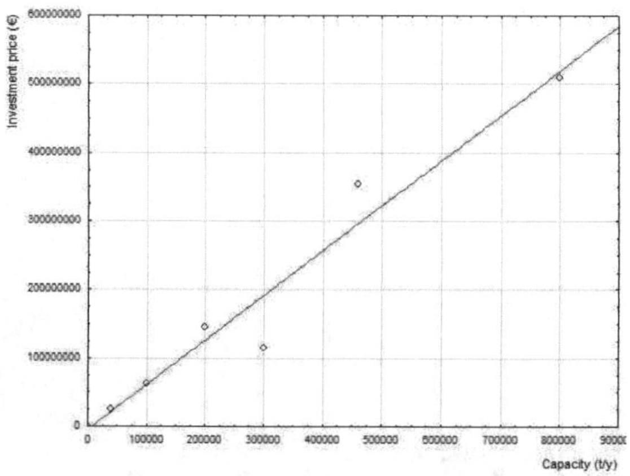

Figure. 3 The dependence of investment price on the capacity

The dependence of specific investments on the capacity is according to the polynomial function:

$$y_1 = -1.5458 \cdot 10^{-9}x^2 + 0.0016x + 317.6361 \tag{1}$$

where is: y_1 – specific investments (€/t) and x – capacity (t/y). The dependence of investment price on the capacityy is shown in Figure 3. The dependence of investment price on the capacity is according to the linear function:

$$y_2 = 654.12x - 4.54 \cdot 10^6 \tag{2}$$

where is: y_2 – investment price (€) and x – capacity (t/y).

2.2. Operating costs and plant maintenance costs

Operating costs include the cost of electricity to drive circulating pumps that serve to maintain the forced circulation of media in the pipelines. Fixed costs are estimated at 0.5% of the investment price of the network, while electricity consumption is calculated from 8 kWhe/MWht for decentralized systems to 12 kWhe/MWht for centralized systems. The water loss in the system is compensated by fresh water. [2]

The next type of costs included in the analysis are operating costs and maintenance costs, and they are presented in absolute form and in the form of a specific value per ton of waste. Depending on the capacity of the plant, the specific value of costs ranges from 46 €/t to 113.20 €/t (Table 3).

Table 3. Overview of operating costs and plant maintenance costs

Capacity (t/y)	Operating and maintenance costs (€)	Specific operating and maintenance costs (€/t)	References
40.000	4.528.021,00	113,20	[3]
100.000	6.427.360,00	64,27	[4]
200.000	9.200.000,00	46,00	[5] [6]
300.000	17.734.043,00	59,11	[3]

The dependence of operating and maintenance costs on the capacity is shown in Figure 4. The dependence of operating and maintenance costs on the capacity is according to the linear function:

$$y_3 = 52.52x + 8.56 \cdot 10^5 \tag{3}$$

where is: y_3 – operating and maintenance costs (€/y) and x – capacity (t/y). The dependence of specific operating and maintenance costs on the capacity is shown in Figure 5. The dependence of specific operating and maintenance costs on the capacity is according to the logarithmic function:

$$y_4 = 416.00 - bg\ 10(x) \tag{4}$$

where is: y_4 – specific operating and maintenance costs (€/t) and x – capacity (t/y).

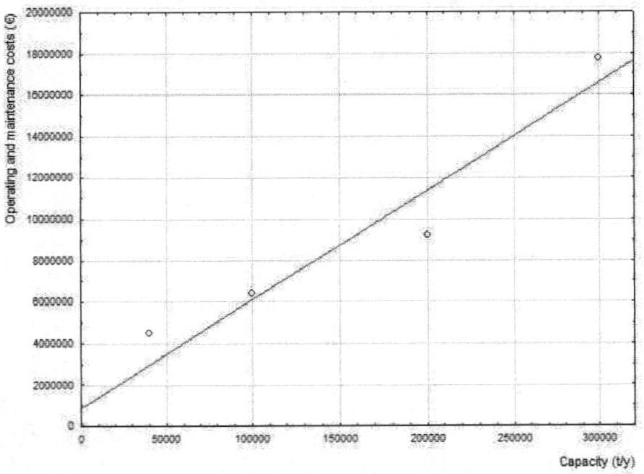

Figure. 4 The dependence of operating and maintenance costs on the capacity

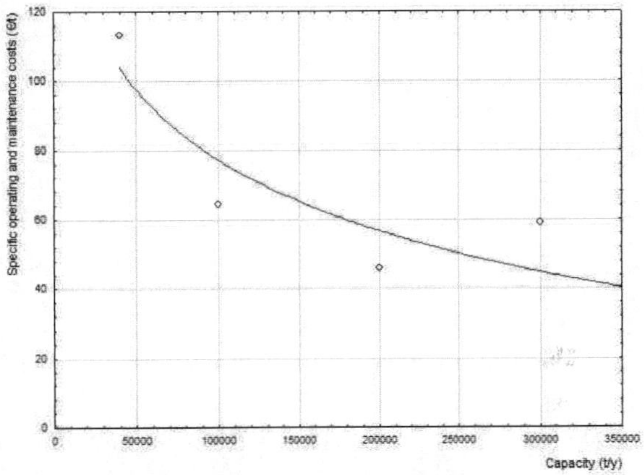

Figure. 5 The dependence of specific operating and maintenance costs on the capacity

2.3. Plant efficiency

In order to evaluate the plant, it is necessary to know the efficiency of energy conversion of fuel energy that it allows. Since the efficiency of each steam turbine plant depends on the steam parameters at the turbine inlet, the energy conversion efficiency increases with the capacity of the plant. [2] The efficiency of the plant is calculated from the annual amount of thermal energy introduced by fuel (waste) and the annual produced quantities of electricity and heat. The efficiency of the plant range from 43.82% to 61.01%, and depends on the capacity of the plant (Table 4).

Table 4. Overview of total plant efficiency

Capacity (t/y)	Year of construction	Total plant efficiency [%]	References
100.000	2010.	43,82	[4]
160.000	1999.	45,00	[7]
460.000	2007.	50,56	[7]
800.000	2004.	61,01	[7]

The dependence of total plant efficiency on the capacity is according to the linear function:

$$y_5 = 2.42 \cdot 10^{-5}x + 40{,}89 \tag{5}$$

where is: y_5 – investment cost (%) and x – capacity (t/y).

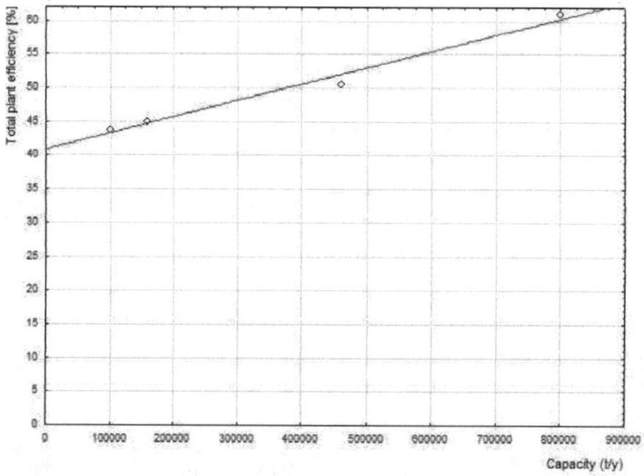

Figure. 6 The dependence of total plant efficiency on the capacity

2.4. Total cost for a Mixed Municipal Waste Incinerators for Waste-to-energy (WtE) in small town

This calculation takes into account the amount of waste for town of 45,000 inhabitants. In the town of Varaždin (a town in the north of Croatia) about 9500 t/y of mixed municipal waste is generated, but in Varaždin, mixed municipal waste was collected, mechanically treated (process D14), baled and stored (process D15) from 2005. until 2013. In that period, about 100.000 t of bales were generated [8]. In this paper, the total costs f the plant with an annual waste capacity of 15.000 t/y of mixed municipal waste is calculated.

According to (1) specific investment (y1) is: 341,28 €/t. According to (2) investment price (y2) is: 5.271.800 €. According to (3) operating and maintenance costs (y3) is: 1.643.800,00 €/y. According to (4) specific operating and maintenance costs (y4) is: 411,82 €/t. According to (5) total plant efficiency (y5) is: 41,253%. For the cogeneration relationship between the production of electricity and heat, the average value from previous plants is used - 0.53. With this ratio and the total efficiency of the plant, the thermal efficiency is 30.18% and the electrical efficiency 15.87% [2]. With the total annual number of hours and the lower heating value of fuel of 10 MJ/kg, data were obtained for the economic analysis of the plant (Table 5)

Table 5. Obtained for the economic analysis

Capacity (t/y)	15.000,00
Specific investment (€/t)	341,28
Specific operating and maintenance costs (€/t)	411,82
Total plant efficiency (%)	41,25
The ratio of electricity and heat	0,53
Thermal efficiency (η_t) (%)	30,18

Electrical efficiency (η_e) (%)	15,87
Total number of working hours per year (h/g)	6.000
Lower calorific value of waste (H_d) (MJ/kg)	10
Heat capacity of the furnace (MW_h)	85,19
Total power (MW_{tot})	39,23
Thermal power (MW_t)	25,71
Electrical efficiency (MW_e)	13,52
Energy obtained (MWh/y)	235.380,00
Thermal energy produced (MWh/y)	154.260,00
Electrical energy produced (MWh/y)	81.120,00
Investment price (€)	5.271.800,00
Operating and maintenance costs (€/y)	1.643.800,00

3. Calculation of Internal rate of return (IRR)

The internal rate of return is the discount rate that reduces the net present value to zero. [35] Most of the revenue will be generated through the gate fee and the sale of heat and electricity. Gate fee varies within the EU (from 46 €/t to 174 €/t [9]), in this paper the average value of 110 €/t is used. The price in the case of Varaždin is calculated as the combined average price of heat and electricity. In addition to revenues from the fee for waste disposal and the sale of electricity and heat, it is also possible to generate revenues from the sale of secondary metals obtained from ash separation. The amount of iron is 0,024 t/tw, and the amount of aluminium is 0,004 t/tw [4]. Prices from the purchase of secondary raw materials were taken from local purchasers. Revenues and expenses of the plant are in Table 6.

Table 6. Revenues and expenses of the plant

Capacity (t/y)	15.000,00
Gate fee (€/t)	110,00
Revenue from waste disposal fees (€/y)	1.650.000,00
The amount of iron (t/t_w)	0,024
The amount of aluminium (t/t_w)	0,004
Quantity of iron (t/y)	360,00
Quantity of aluminium (t/y)	60,00
Secondary iron price (€/t)	105,26
Secondary aluminium price (€/t)	657,89
Revenue from the sale of secondary iron (€/y)	37.893,60
Revenue from the sale of secondary aluminium (€/y)	39.473,40
Sales price of electricity (€/kWh)	0,0770
Sales price of thermal energy (€/kWh)	0,0281
Thermal energy produced (MWh/y)	154.260,00
Electrical energy produced (MWh/y)	81.120,00
Revenue from electrical energy sales (€/year)	6.246.240,00
Revenue from thermal energy sales (€/year)	4.334.706,00

4. Conclusion

Due to profitability and rentability factors, including the waste management internal return rate, there is a wider picture to consider. Good waste management yields a cleaner and more pleasurable environment as the positive externality, which brings higher property prices [10] and directly influences the national wealth. Furthermore, such an environment increases the desirability for the property buyers and brings new taxpayers to the municipality that increases local fiscal capacity. As presented above, Mixed Municipal Waste Incinerators are not a quick fix for the waste management problems and their cost is offset only in the long term, longer than a few election cycles. Municipality, though, needs to charge their services at a feasible minimum in order to maintain the highest possible living standard level. For that reason, services are often charged less than their real economic cost and the profitability in the public service is not a decisive issue [11] because municipalities expect to benefit through positive externalities as a consequence of local policies [12]. In that sense, Mixed Municipal Waste Incinerators cost-effectiveness model presented in this paper could offer help in decision-making processes and is developed for such a purpose.

References

[1] Gotal Dmitrovic L, Dusak V, Anic Vucinic A (2015) The Development of Conceptual, Mathematical and System Dynamics Model for Food Industry Wastewater Purifying System, Journal of Information and Organizational Sciences - JIOS vol. 39 No 2

[2] Tomić T. (2014) Analiza isplativosti centraliziranog toplinskog sustava na biomasu i komunalni otpad, Zagreb

[3] European Commission: Reference Document on the Best Available Techniques for Waste Incineration; 2006; http://eippcb.jrc.es/pages/FActivities.html

[4] Schneider D R, Lončar D, Bogdan Ž (2010) Cost Analysis of Waste-to-Energy Plant, Strojarstvo (0562-1887) Vol. 52, No 3; 369-378

[5] Friotherm AG: Waste-to-Energy Plant Sysav Malmo, enhancing the overall Energy Efficiency with 2 UnitopR 28C heat pump units http://www.friotherm.com/webautor-data/41/sysave006_uk.pdf

[6] ENEM - Epem S.A. Environmental planning and management: Waste Control: Database of Waste Management Technologies http://www.epem.gr/waste-c-control/database/html/costdata-00.htm#Incineration

[7] Wood S, Fanning M, Venn M, Whiting K (2013) Review of state-of-the-art waste-to-energy technologies, Stage two: Case studies, Department of Environment and Conservation: Waste Management Branch; Perth, Australia

[8] Gotal Dmitrovic, L (2020) Reduction of Dissolved Organic Carbon in Baled, Mixed, Municipal Waste Through 12 Years, IOP Conference Series: Earth and Environmental Science (EES), Vol. 563

[9] Watkins, E, Hogg, D, Mitsios, A, Mudgal, S, Neubauer, A, Reisinger, H, Troeltzsch, J, Acoleyen, M.V. (2012) Use of economic instruments and waste management performances; Bio Intelligence Service S.A.S.; Paris, France

[10] Boyle, M.A. and Kiel, K.A. (2001) A survey of house price hedonic studies of the impact of environmental externalities, Journal of Real Estate Literature, Vol. 9 No. 2, pp. 117-144

[11] Scutaru, C., (2009) Marketing for public sector, Calitatea Vieii, 1-2: 63. Bucarest

[12] Schmitz A., Moss C.B., Schmitz T.G., Furtan H.W. and H.C. Schmitz (2010) Agricultural Policy, Agribusiness and Rent-Seeking Behaviour, 2nd Edition. Toronto: University of Toronto Press

IOP Conf. Series: Materials Science and Engineering 1150 (2021) 012003 doi:10.1088/1757-899X/1150/1/012003

Application of Generalized Models for Identification of Viscoelastic Behaviour on Hysteretic Double Peak Materials

F Carputo, V M Arricale, R Brancati, A Maiorano and L Mosconi

University of Naples Federico II

E-mail: antonio.maiorano@unina.it

Abstract. This paper investigates the capability of the fraction derivative Maxwell model to reproduce the mechanical behaviour of advanced viscoelastic materials, that are designed to be able to dissipate energy in an extended range of frequency than traditional viscoelastic materials and present a more complex shape of strain-stress frequency response. Fraction derivative models, compared to traditionally used generalized Maxwell model, can reproduce the material behaviour with a smaller number of parameters, reducing the utilization complexity and the computational burden.

1. Introduction

Viscoelastic materials present a time dependent stress-strain relationship, that must be well known to get a correct design of the component. The application field is extremely wide, starting from vibration control problems in industrial machines [1], vehicle engines, aerospace structures and problems in civil engineering concerning the protection of tall buildings from earthquakes or wind [2]. In human body are also common viscoelastic materials, like the skin tissue [3] or the human spine disks [4].

Tire compound performance is also mainly determined from its viscoelasticity properties, consequently, the dynamic behaviour of the entire vehicle that concerns fundamental aspects like safety and performance [5]. In particular, the main aspects about tire performance are the friction with road surface, rolling resistance and wear [6].

In order to evaluate the friction between rubbery material and substrate it is possible to use different test benches such as the British Pendulum Evo [7]. Instead, the crucial information for a rough surface is its power spectrum [8].

To characterize viscoelastic materials are needed both experimental testing and analytical modelling. Those models will often study the behaviour of the material in the structure where is used, and their rheological properties must be considered with the component's geometric parameters.

These characteristics will depend on the temperature and on the excitation frequency. In this context, simulation tools are needed to get correlation with experimental testing environment and the application one, reducing the uncertainties during the various stages of the material development.

In the past years, studies showed how the mechanical behaviour of viscoelastic materials, measured with dynamic mechanical analysis (DMA) [9] can be efficiently reproduced using a Maxwell model with fraction derivative [10]. Fractional derivative Maxwell model (FDMG), compared to a Generalized Maxwell model (GM), can reproduce the experimental results using a lower number of parameters.

The cited technique was successfully applied on experimental data relative to materials characterized by the traditional viscoelastic behaviour, where the phase shift between stress and strain, that is strictly correlated to energy dissipation, present a single peak in all the frequency range.

In this paper is shown that the same technique is suitable even for advanced materials that are designed to present two peaks of stress-strain phase shift in the frequency range. This feature consent to extend significantly the working range where the component purpose is the energy dissipation like the protection of buildings from earthquakes or in high performance motorsport tires.

2. Rheological models

Viscoelastic material's constitutive law is described with the more convenient and common differential approach, where the rheological properties are described by a combination of ideal mechanical elements, springs and dashpots. The constitutive equation represents the relationship by stress and strain, described by a differential equation.

A way to judge the model goodness is the number of parameters needed to describe the experimental data. If many parameters are needed, data correlation will be more difficult and model reliability can be lower, so the best model will be that one that can accurately represent the material behaviour with a lower number of parameters.

The most famous models are Maxwell model, Zener model and Kelvin-Voigt model. Those are able to reproduce experimental data only on low frequency or high frequency ranges, so, generalized models have been created to describe the viscoelastic behaviour in a larger range of frequencies. Generalized Maxwell model (GM) and generalized Kelvin-Voigt model (GKV) are able to reproduce accurately the material characteristics only if a larger number of spring and dashpots elements is used, this will raise the complexity of the differential equation that must be solved to simulate the dynamic behaviour of the components [11].

The described issues with generalized models have been overcome by fractional models, that raised in popularity because are able to describe accurately he viscoelastic characteristics in a wide frequency range using a small number of parameters [12][13][14].

Generalized Maxwell model and Fractional derivative Maxwell model are made by a spring in parallel with Maxwell elements or fractional Maxwell elements, as defined by Koeller [15]. Maxwell elements are made by a spring and a dashpot in series. In a fractional Maxwell element, a dashpot is replaced by a spring-pot element, represented in Fig. 1. A spring-pot element generates a force proportional to the deformation derivative order included between 0 and 1.

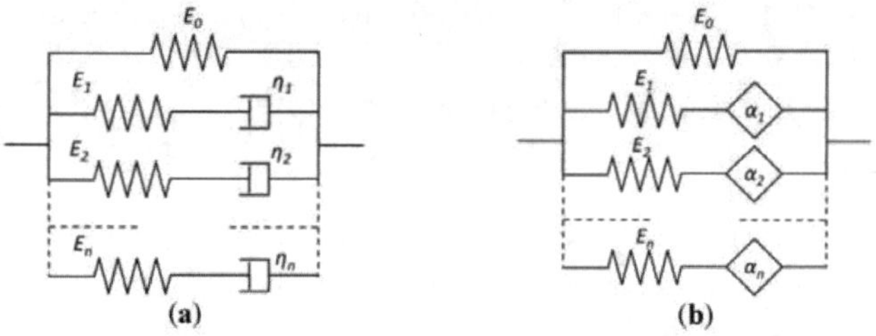

Figure. 1 Rheological models: (a) Generalized Maxwell, (b) Fractal derivative.

The mathematical description of the GM model starts from the following equation of the complex moduli E^*, which represents the stress-strain relationship at a certain frequency.

$$E^*(i\omega) = E_0 + \sum_{k=1}^{N} \frac{i\omega E_k \eta_k}{E_k + i\omega E_k \eta_k} \tag{1}$$

Where E_k and η_k are the springs stiffness and the dashpot viscosity parameters, showed in figure 1a, and omega is the angular frequency.

The same complex moduli can be described with an equivalent equation, demonstrated by Renaud et al. [16] where instead of parameters relative to elements properties, appears poles and zeros, helping

IOP Conf. Series: Materials Science and Engineering 1150 (2021) 012003 doi:10.1088/1757-899X/1150/1/012003

the utilization of the formula as those parameters are related to the shape of the stress-strain transfer function gain and delay graphs.

$$E^*(i\omega) = E_0 + \prod_{k=1}^{N} \frac{1+\left(\frac{i\omega}{\omega_{z,k}}\right)}{1+\left(\frac{i\omega}{\omega_{p,k}}\right)} \tag{2}$$

Where ω_p and ω_z are poles and zeros.

Going to Fractional Derivative Generalized models, complex moduli can be similarly expressed as:

$$E^*(i\omega) = E_0 + \sum_{k=1}^{N} \frac{(i\omega)^{\alpha_k} E_k \eta_k}{E_k + (i\omega)^{\alpha_k} E_k \eta_k} \tag{3}$$

Where α_k is introduced, which represents the fractional derivative order and must be included between 0 and 1. As the GM model, FDGM can be expressed with poles and zeros:

$$E^*(i\omega) = E_0 + \prod_{k=1}^{N} \frac{1+\left(\frac{i\omega}{\omega_{z,k}}\right)^{\alpha_k}}{1+\left(\frac{i\omega}{\omega_{p,k}}\right)^{\alpha_k}} \tag{4}$$

3. Material under investigation

Polymer experimental data obtained by DMA characterization is provided. This test is a standard way to measure the complex modulus is carried out by deforming the specimen imposing a sinusoidal stress or strain with constant frequency and amplitude. This test is repeated at different frequencies and temperatures in order to explore all the component working range. Temperature must be considered because has influence on the viscoelastic characteristics comparable to the one induced by an excitation frequency change, and an equivalency can be found through the time-temperature superposition principle, also called Williams – Landel – Ferry law [17].

Most of materials present a single peak of $\tan(\delta)$ in function with excitation frequency, the phase shift between real and imaginary part of the complex modulus E^*, $\tan(\delta)$ is strictly correlated with energy dissipation and because of the WLF law this peak can heavily translate in frequency when the component is used at different temperatures, this can be critical in components that are designed to work in large temperature ranges such as motorsport tires [18].

To overcome this problem, compound designers created materials that present a double peak of $\tan(\delta)$, in order to extend the working range both in frequency and temperature.

The material examined in this paper was measured at temperature 120°C, it is characterized by glassy transition temperature T_g = -5°C and a dynamic modulus at low frequencies E_0 = 25 MPa.

As mentioned before, this is an advanced material that presents a double $\tan(\delta)$ peak, unlike normal polymers that presents a single peak.

4. Identification of viscoelastic behaviour

In order to identify the parameters, the pole-zero identification has been considered.

Such formula refers to the achievement of the poles and zeros of a linear or linearized system described by its frequency response. Usually optimization techniques are used to adapt a given frequency response of the linear system towards a transfer function defined as the ratio of two polynomials [19].

In this paper the identification of poles and zeros has been carried out using a constrained procedure that uses Matlab "fminmax" function which minimizes a non-linear error function W of n real variables representing the Zeros and Poles, to fit both modulus and phase at the same time:

$$W = (1 - r) \cdot Err_E + r \cdot Err_{\tan(\delta)} \quad (5)$$

The function W is defined as the weighted sum, where r is the weight factor, of the normalized root mean square error (NRMSE) calculated for the storage modulus and the loss tangent:

$$Err_{\tan(\delta)} = \frac{\sqrt{\sum_{i=1}^{N}(\tan(\delta) - \tan(\delta)_{model})^2}}{mean(\tan(\delta))} \quad (6)$$

$$Err_E = \frac{\sqrt{\sum_{i=1}^{N}(E - E_{model})^2}}{mean(E)} \quad (7)$$

The constrain imposed to the function W, according to the third property demonstrated by Bland [20], requires that the poles p and zeros z be alternate $z_1 < p_1 < z_2 < p_2 \dots < z_i < p_i$.

For the FDGM model described in Fig. 1, the minimum number of parameters required to describe the viscoelastic behaviour of the material must be sought. In this regard, the identification procedure consists in increasing the number of elements for each identification until the fitting of experimental data was not satisfactory. The quantitative comparison of the goodness of the identification in describing the behaviour of the material is made by evaluating the NRMSE for storage modulus and loss tangent quantities as defined in the equation (6).

In the Fig. 2 and 3 are shown the comparison between the experimental data in terms of storage modulus (a) and loss tangent (b) with the results obtained with the FDGM models composed respectively by 3 and 6 elements plus a spring in parallel (E_0 in Fig. 1), while in table 1 and 2 the relative calculated NRMSEs are reported.

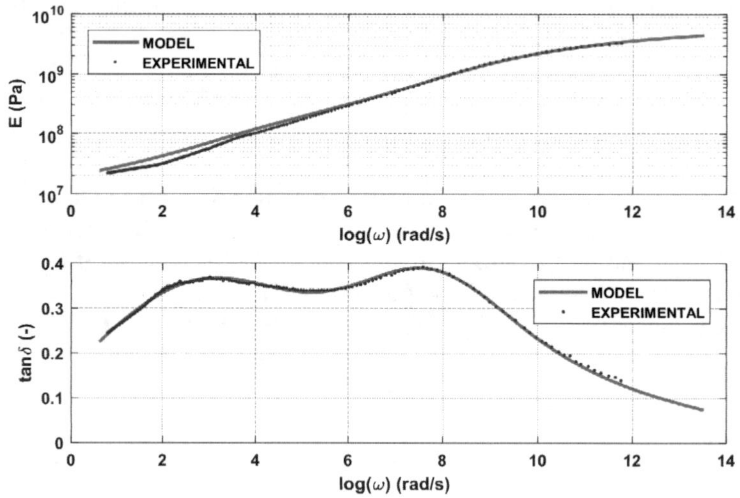

Figure. 2 Storage modulus (a) and loss factor (b): experimental data (dashed line) vs. FDGM model with 3 elements.

Table 1. NRMSE for FDGM models with 3 elements with r = 0.8

Err $\boldsymbol{E'}$	0.0555
Err $\boldsymbol{\tan(\delta)}$	0.011
Err tot	0.0199

It is possible to observe that the FDGM model is able to give a good representation of the shapes of the curves already with a model with only 3 elements that is also consistent with the NRMSE values. By increasing the number of elements of the model from 3 to 6 elements, it improves the adaptation of the experimental data even if in a very slight way, because the shape of the curve remains almost identical but improves the NRMSE. From the simple comparison of the normalized errors, it is generally observed a good performance of both identifications adopted in describing the behaviour of the real compounds.

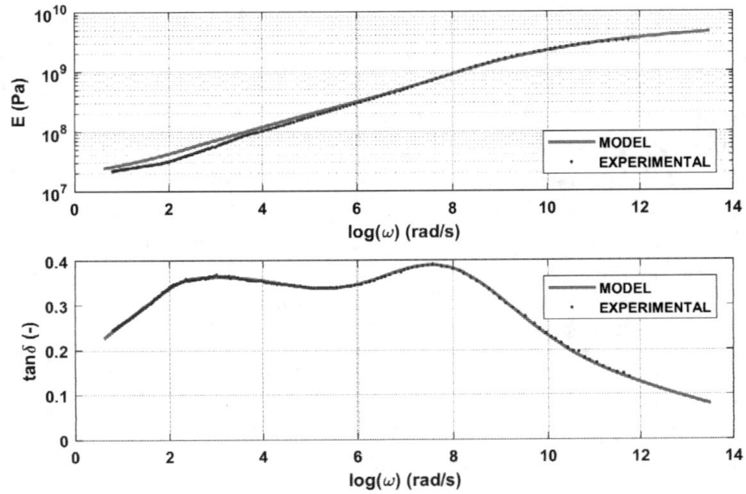

Figure. 3 Storage modulus (a) and loss factor (b): experimental data (dashed line) vs. FDGM model with 6 elements

Table 2. NRMSE for FDGM models with 6 elements with r = 0.8

Err $\boldsymbol{E'}$	0.0586
Err $\boldsymbol{tan}\ (\boldsymbol{\delta})$	0.0063
Err tot	0.0167

Using a zero-pole formulation has the advantage of having a range of parameters to identify (poles and zeros) defined and limited by the frequency range of the experimental curves, thus greatly facilitating the efficiency of the optimization algorithms.

5. Conclusions

In this paper a study on a possible adoption of generalized models for the reproduction of the behaviour of viscoelastic materials is presented, in particular for materials with a double $\tan(\delta)$ peak.

The FDGM model has been expressed in terms of a Pole-Zero formulation. In addition, a procedure for the identification of poles and zeros has been developed and adopted in order to assess the goodness of the FDGM model to reproduce the characteristic curves of the materials examined. This goodness was quantified by evaluating the NRMSE between the values obtained from the model and the experimental data acquired through DMA technique.

The analysis has shown that the FDGM model is able to adapt to the viscoelastic behaviour of the compound in the entire frequency range in which it is characterized experimentally, it also shows that the FDGM model adapts well even with a limited number of elements and therefore of parameters to be identified, reducing the computational burden of the identification algorithms.

References

[1] Park S W *Analytical modeling of viscoelastic dampers for structural and vibration control.* International Journal of Solids and structures (Amsterdam: Elsevier) 38.44-45 (2001): 8065-8092
[2] Singh M P and Moreschi L M *Optimal placement of dampers for passive response control.* Earthquake engineering & structural dynamics (Hoboken, Wiley) 31.4 (2002): 955-976
[3] Zhang W, Chen Y H, and Kassab G S. *A rate-insensitive linear viscoelastic model for soft tissues.*

Biomaterials (Amsterdam: Elsevier) 28.24 (2007): 3579-3586
[4] Newell N, et al. *Biomechanics of the human intervertebral disc: a review of testing techniques and results.* Journal of the mechanical behaviour of biomedical materials (Amsterdam: Elsevier) 69 (2017): 420-434.
[5] Romano L, Sakhnevych A, Strano S and Timpone F 2019 *A novel brush-model with flexible carcass for transient interactions.* Meccanica (New York City: Springer) , 54(10), 1663-1679
[6] Sharifzadeh M, Senatore A, Farnam A, Akbari A and Timpone F (2018). *A real-time approach to robust identification of tyre–road friction characteristics on mixed-μ roads.* Vehicle system dynamics (Oxfordshire: Taylor & Francis)
[7] Ciaravola V, Farroni F, Fortunato G, Russo M, Russo R, Sakhnevych A and Timpone F 2017 *An Evolved Version of the British Pendulum Tester for the Experimental Investigation of Contact Between Tire Tread and Rough Surfaces.* Engineering Letters, 25(1).
[8] Genovese A, Farroni F, Papangelo A and Ciavarella M 2019 *A Discussion on Present Theories of Rubber Friction, with Particular Reference to Different Possible Choices of Arbitrary Roughness Cutoff Parameters.* Lubricants (Basel, MDPI), 7(10), 85.
[9] Menard K P and Menard N 2017 *Dynamic Mechanical Analysis* in Encycl. Anal. Chem. (Chichester: John Wiley & Sons), UK, pp. 1–25.
[10] Genovese, A, Carputo, F, Maiorano, A, Timpone, F, Farroni, F, Sakhnevych 2020 A. *Study on the Generalized Formulations with the Aim to Reproduce the Viscoelastic Dynamic Behaviour of Polymers.* Appl. Sci. (Basel, MDPI) 10, 2321.
[11] Palmeri, A, Ricciardelli, F, De Luca, A, & Muscolino, G. (2003). *State space formulation for linear viscoelastic dynamic systems with memory.* Journal of Engineering Mechanics, 129(7), 715-724.
[12] Bagley, R L, & Torvik, P J (1983). *Fractional calculus-a different approach to the analysis of viscoelastically damped structures.* AIAA journal, 21(5), 741-748.
[13] Pritz, T. (2003). *Five-parameter fractional derivative model for polymeric damping materials.* Journal of Sound and Vibration, 265(5), 935-952.
[14] Sasso M, Palmieri G, and Amodio D 2001 Mech. Time-Dependent Mater. 15, 367
[15] Koeller R C 1984 J *Appl. Mech. Trans.* ASME 51, 299.
[16] Renaud, F, Dion, J L, Chevallier, G, Tawfiq, I, & Lemaire, R (2011). *A new identification method of viscoelastic behaviour: Application to the generalized Maxwell model.* Mechanical Systems and Signal Processing, 25(3), 991-1010.
[17] Williams, M L, Landel, R. F, & Ferry, J D (1955). *The temperature dependence of relaxation mechanisms in amorphous polymers and other glass-forming liquids.* Journal of the American Chemical society, 77(14), 3701-3707..
[18] Allouis, C, Farroni, F, Sakhnevych, A, & Timpone, F 2016. *Tire thermal characterization: test procedure and model parameters evaluation.* In Proceedings of the World Congress on engineering (Vol. 2).
[19] Pintelon R and Schoukens J 2012 *System Identification: A Frequency Domain Approach*, Second Edition (Wiley, 2012)
[20] Bland D R 2019 *The Theory of Linear Viscoelasticity* (Pergamon Press, London, 2019).

Influence of Compacting Pressure on the Dielectric Properties of La-modified Bismuth Ferrite Multiferroics Prepared by Rapid Liquid-phase Sintering Method

Chunrui Song[1,a], N.A. Liedienov[1,2,b] *, A.V. Pashchenko[1,2,3,c], D.D. Tatarchuk[4,d], Y.V. Didenko[4,e] and G.G. Levchenko[1,2,f]

[1]State Key Laboratory of Superhard Materials, International Center of Future Science, Jilin University, 130012 Changchun, China
[2]Donetsk Institute for Physics and Engineering named after O.O. Galkin, NASU, 03028 Kyiv, Ukraine
[3]Institute of Magnetism NASU and MESU, 03142 Kyiv, Ukraine
[4]National Technical University of Ukraine "Igor Sikorsky KPI", 03056 Kyiv, Ukraine

E-mail: Liedienov@nas.gov.ua

Abstract. Structure, perovskite phase formation, chemical and phase composition, microstructure, morphology and dielectric properties of the single-phase lanthanum-modified bismuth ferrite multiferroic ceramics obtained by the rapid liquid-phase sintering method under different compacting pressures have been investigated using X-ray diffraction, scanning electron microscopy, energy-dispersive X-ray spectroscopy, thermogravimetry/differential thermal analysis and dielectric spectroscopy methods. The temperature and time for an express preparation of the single-phase $Bi_{0.9}La_{0.1}FeO_3$ ceramics with formation of a perovskite structure have been found out. With an increase in the compacting pressure P from 100 to 1100 MPa, the dielectric constant ε' changes by ~25% and ~80% in the low-frequency and ultra-high frequency ranges, respectively. In the low-frequency range, all samples depending on the pressure P demonstrate excellent dielectric behavior with a stable dielectric constant and quite low dielectric loss that make these samples perspective for their possible practical application in microelectronic and spintronic.

1. Introduction

The multiferroics with a perovskite structure demonstrate at least two ferro-order states which not only have a variety of single ferrous properties, but also have some new effects through the coupling and compound synergistic effect of ferrous properties, which greatly broadens the application range of these metal-oxide materials [1–6]. The $Bi_{0.9}La_{0.1}FeO_3$ as one of multiferroics show high values of dielectric constant [7,8]. However, during synthesis of this composition, the minor $Bi_{25}FeO_{40}$ and $Bi_2Fe_4O_9$ phases can be appeared which leads to leakage current and deterioration of the properties. There are a lot of different preparation methods to exclude these second phases and to obtain single-phase multiferroic material, but all of them have some advantages and drawbacks. According to the experiment [7,8], the rapid liquid-phase sintering (RLS) method is most perspective and promising method to avoid these minor phases and to decrease structure defects effectively. In this paper, using the RLS method as a basis, the optimal conditions for an express obtaining single-phase $Bi_{0.9}La_{0.1}FeO_3$ ceramics under different compacting pressures P = 100–1100 MPa of initial mixture powder of precursors were determined and the influence of the P on the dielectric properties in a wide frequency range was studied.

2. Experimental section

The $Bi_{0.9}La_{0.1}FeO_3$ ceramic samples were synthesized using the RLS method [9] at 850 °C (480 s) in air under different compacting pressures P = 100–1100 MPa of initial La_2O_3, Bi_2O_3 and Fe_2O_3 precursors. The formation of a perovskite structure in the obtained samples was detected by thermogravimetry/differential thermal analysis (TG/DTA) using Linseis STA PT 1600 within temperature range 25–900 °C. Crystal structure and lattice parameters were determined using high-resolution PANalitical X-Pert PRO MRD diffractometer in $CuK_{\alpha 1}$-radiation. Microstructure, phase and chemical composition, as well as grain size were tested using FEI MAGELLAN 400 Scanning Electron Microscope (SEM) with energy-dispersive X-ray spectroscopy (EDS) mode. The frequency dependences of the dielectric constant ε' in the low-frequency (LF) 0.1–10^6 Hz range and in the ultra-high frequency (UHF) 8–12 GHz range were measured using dielectric spectroscopy method [7].

3. Results and discussion

According to TG/DTA data for the Bi_2O_3–La_2O_3–Fe_2O_3 precursors, at heating mode (30°C·min^{-1}), several endothermic and exothermic effects in the DTA (red curve) are observed from 400 to 800 °C (see Fig. 1(a)). In the range of 400–500 °C, endothermic effects may be due to the decomposition of carbonate species and surface hydroxyl groups. In the range of 600–800 °C, exothermic effects are due to the formation of minor $Bi_{25}FeO_{40}$ and $Bi_2Fe_4O_9$ phases [9]. At cooling mode, a small exothermic effect near 850 °C is noted in the DTA (blue curve) that is due to the formation process of perovskite phase [9]. At the selected temperature of 850 °C during an isothermal process, an exothermic peak appears after ~ 5 min (see Fig. 1(b)). Simultaneously, the total mass loss is almost unchanged (see Fig. 1(b)).

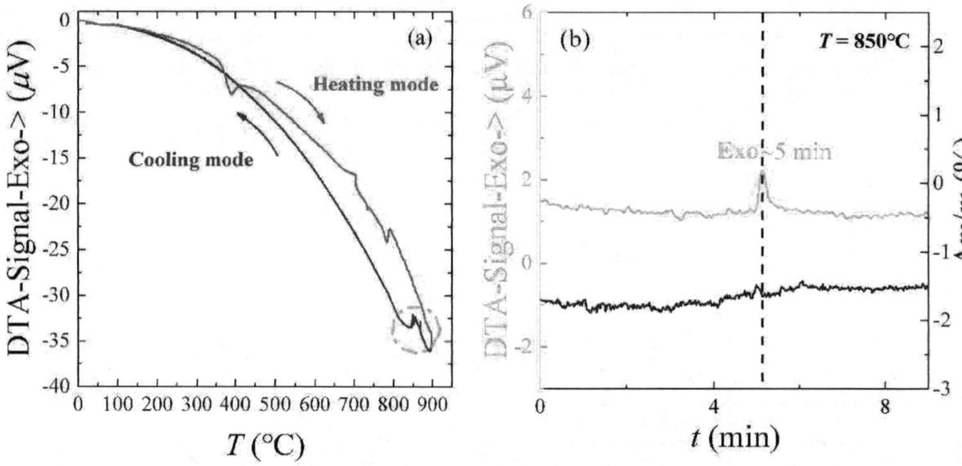

Figure 1. DTA curves recorded in heating (red curve) and cooling (blue curve) modes for the Bi_2O_3–La_2O_3–Fe_2O_3 precursor (a) and the time dependences of DTA (green curve) and total mass loss change (navy curve) at the selected temperature of 850 °C (b).

According to X-ray diffraction (XRD) data, the formation of the perovskite structure for the mixture of precursors before sintering is not observed and their stoichiometric ratio is confirmed (see Fig. 2(a)).

After synthesis, the $Bi_{0.9}La_{0.1}FeO_3$ ceramics become single-phase with a rhombohedral *R3c* perovskite structure. The symmetry and lattice parameters of the $Bi_{0.9}La_{0.1}FeO_3$ are almost unchanged under compacting pressure *P*.

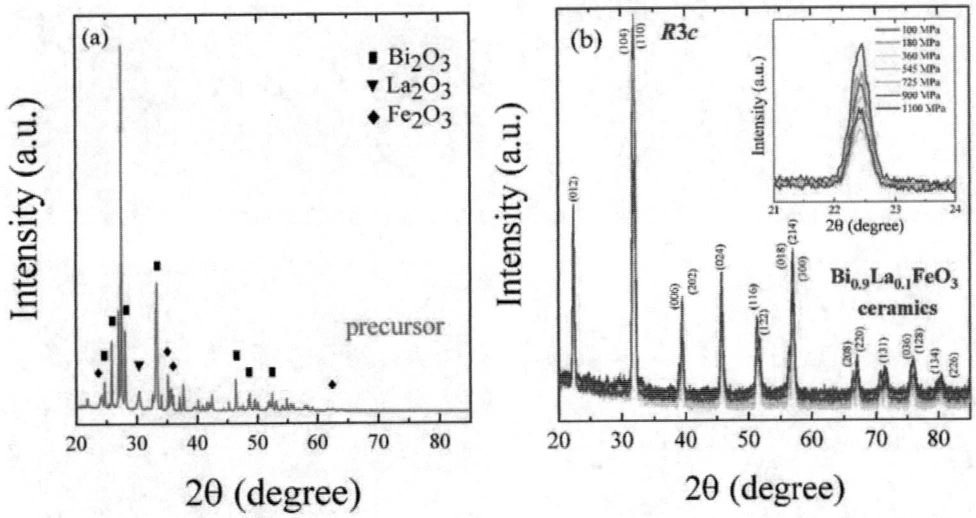

Figure 2. XRD patterns for the initial mixture of precursor (a) and for the $Bi_{0.9}La_{0.1}FeO_3$ ceramics obtained under different compacting pressures *P*.

According to SEM and EDS data, the stoichiometric Bi_2O_3–La_2O_3–Fe_2O_3 oxides were mixed well (see Fig. 3(a,b)). All $Bi_{0.9}La_{0.1}FeO_3$ ceramic samples show a well-formed microstructure with clear boundaries of grains, which weakly depends on the *P* and increases from 200 to 300 nm (see Fig. 3(c,e,g)). Stoichiometric ratio of precursors and synthesized $Bi_{0.9}La_{0.1}FeO_3$ ceramics (approximately 1:1:3) was additionally confirmed by EDS data (see Fig. 3(d,f,h)).

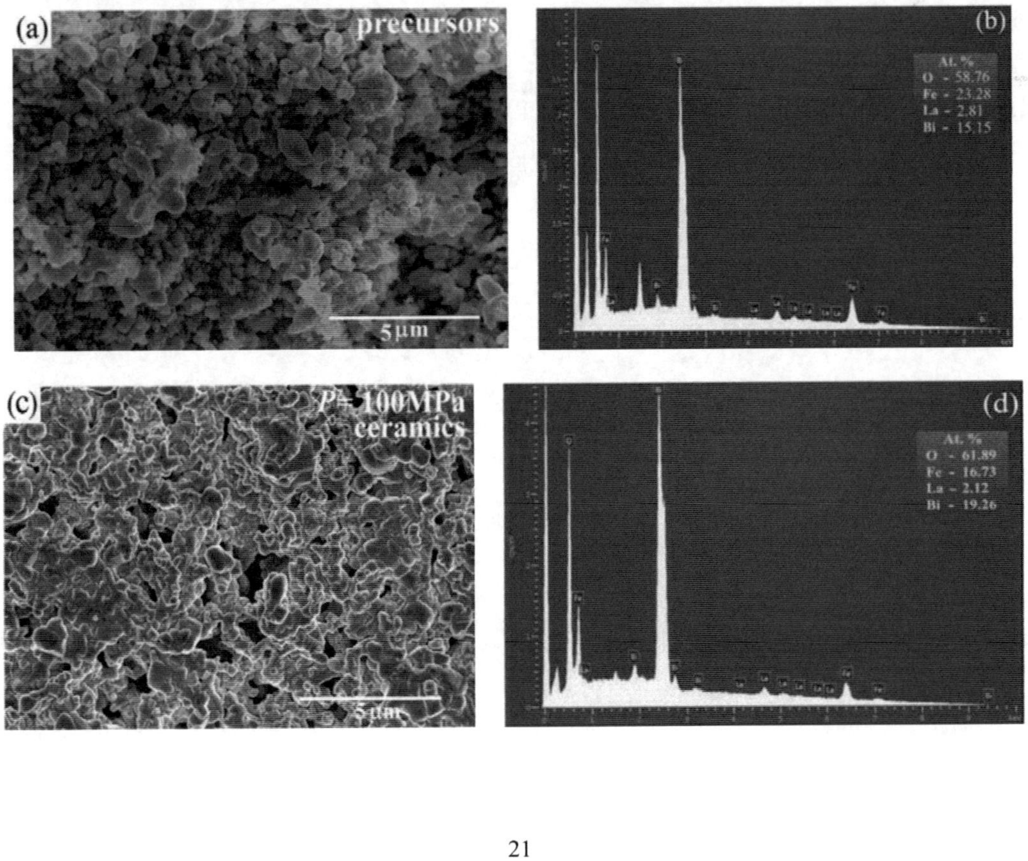

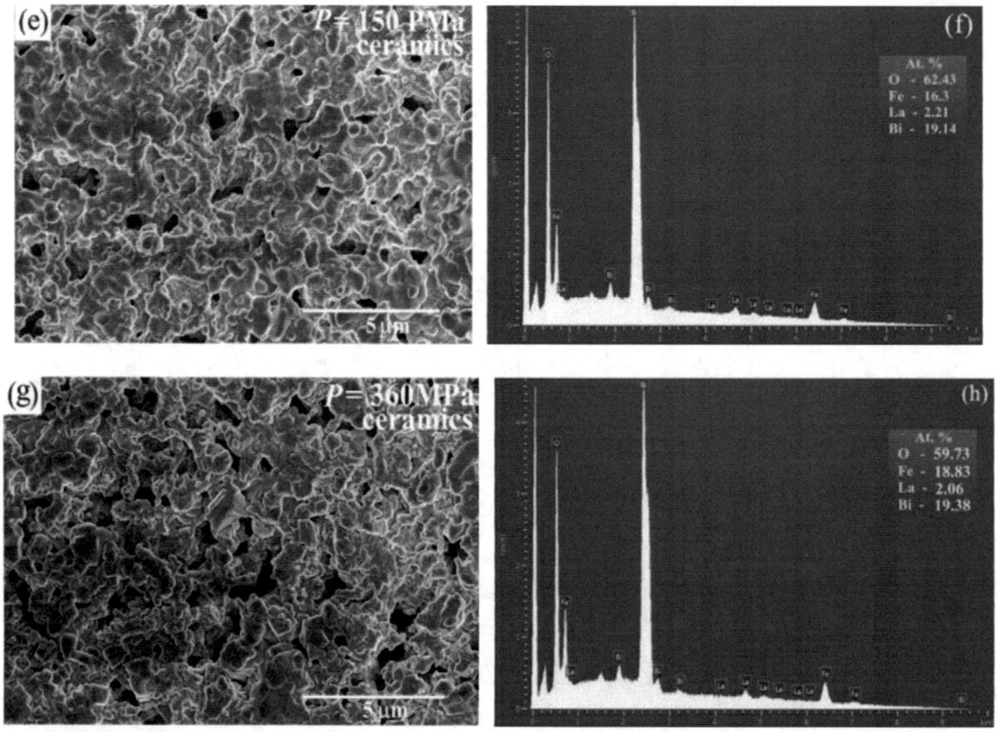

Figure 3. SEM (a,c,e,g) and EDS (b,d,f,h) images for the Bi_2O_3–La_2O_3–Fe_2O_3 oxides and $Bi_{0.9}La_{0.1}FeO_3$ ceramics.

According to the dielectric spectroscopy data (see Fig. 4), the dielectric constant ε′ and dielectric loss tangent tanδ strongly depend on the compacting pressure P. In the LF range (see Fig. 4(a)), the $Bi_{0.9}La_{0.1}FeO_3$ ceramics show excellent dielectric behavior with a stable dielectric constant and quite low dielectric loss. With increase in the P from 100 to 1100 MPa, the relative change Δε′/ε′ is changed by ~ -25% in the LF range (see Fig. 4(b)), that can be associated with a hopping charge transfer between Fe^{3+}↔Fe^{2+} ions and accumulation of weakly bound charge carriers on grain boundaries, point defects and other structural inhomogeneities [7,8]. In the UHF range, the Δε′/ε′ is changed dramatically by ~ -80% and may be associated with ionic and electronic elastic polarizations.

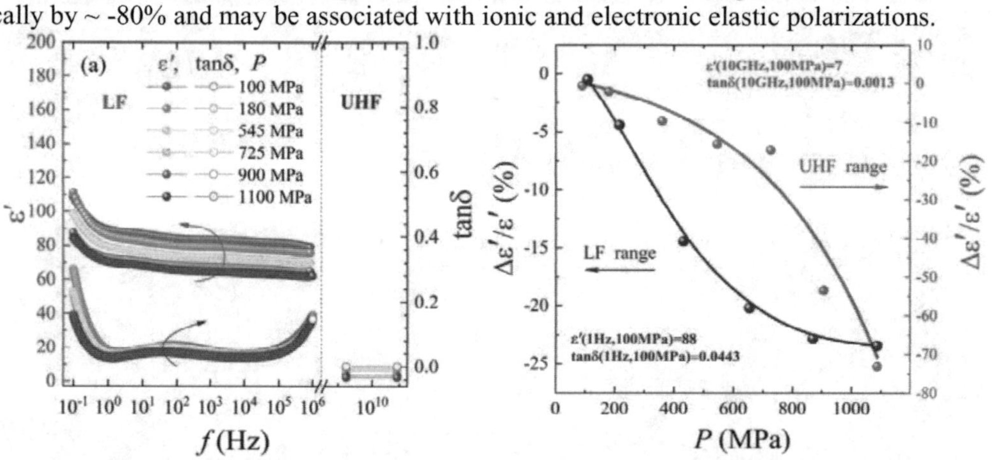

Figure 4. The frequency dependences of ε′(f) and tanδ(f) (a) and compacting pressure dependence of dielectric relative change Δε′/ε′(P) (b) for the $Bi_{0.9}La_{0.1}FeO_3$ ceramics in the LF and UHF ranges.

Conclusions

The optimal conditions 850 °C after ~ 5 min in air for the formation of a perovskite structure in the stoichiometric $Bi_{0.9}La_{0.1}FeO_3$ composition have been determined and confirmed by TG / DTA / XRD / SEM / EDS data. Using rapid-liquid phase sintering method with optimal conditions, the single-phase $Bi_{0.9}La_{0.1}FeO_3$ ceramics under different compacting pressures P have been synthesized. The pressure P from 100 to 1100 MPa can easily change dielectric properties of the $Bi_{0.9}La_{0.1}FeO_3$ multiferroics by ~ 25% in the low-frequency and by ~ 80% in the ultra-high frequency ranges. Varying compacting pressure P at the sintering process, the dielectric constant of the $Bi_{0.9}La_{0.1}FeO_3$ ceramics can be easily controlled in a wide frequency range of 10^{-1}–10^{10} Hz that opens new perspectives for their possible application in microelectronic and spintronic.

Acknowledgments

This work was partially supported by The Thousand Talents Program for Foreign Experts program of China (project WQ20162200339) and Grant of NAS of Ukraine for research laboratories / groups of young scientists of NAS of Ukraine in 2020-2021 (№ 08/01-2021).

References

[1] T. Sluka, A.K. Tagantsev, D. Damjanovic, M. Gureev, N. Setter, Enhanced electromechanical response of ferroelectrics due to charged domain walls, Nature Communications 3 (2012) 748.

[2] P. Maksymovych, Yu.Pu. Morozovska, E.A. Eliseev, Y.-H. Chu, R. Ramesh, A.P. Baddorf, S.V. Kalinin, Tunable Metallic Conductance in Ferroelectric Nanodomains, Nano Letters, 12 (2012) 209–213.

[3] I.I. Makoed, N.A. Liedienov, A.V. Pashchenko, G.G. Levchenko, D.D. Tatarchuk, Y.V. Didenko, A.A. Amirov, G.S. Rimski, K.I. Yanushkevich, Influence of rare-earth doping on the structural and dielectric properties of orthoferrite $La_{0.50}R_{0.50}FeO_3$ ceramics synthesized under high pressure, Journal of Alloys and Compounds, 842 (2020) 155859.

[4] A.A. Amirov, I.I. Makoed, Y.A. Chaudhari, S.T. Bendre, D.M. Yusupov, A.Sh. Asvarov, N.A. Liedienov, A.V. Pashchenko, Magnetocaloric effect in $BiFe_{1-x}Zn_xO_3$ multiferroics, Journal of Superconductivity And Novel Magnetism, 31 (2018) 3283–3288.

[5] I.I. Makoed, A.A. Amirov, N.A. Liedienov, A.V. Pashchenko, K.I. Yanushkevich, Predicted model of magnetocaloric effect in $BiFeO_3$-based multiferroics, Solid State Sciences, 95 (2019) 105920.

[6] I.I. Makoed, A.A. Amirov, N.A. Liedienov, A.V. Pashchenko, K.I. Yanushkevich, D.V. Yakimchuk, E.Yu. Kaniukov, Evolution of Structure and Magnetic Properties in $Eu_xBi_{1-x}FeO_3$ Multiferroics Obtained under High Pressure, Journal of Magnetism and Magnetic Materials, 489 (2019) 165379.

[7] N.A. Liedienov, A.V. Pashchenko, V.A. Turchenko, V.Ya. Sycheva, A.V. Voznyak, V.P. Kladko, A.I. Gudimenko, D.D. Tatarchuk, Yu.V. Didenko, I.V. Fesych, I.I. Makoed, A.T. Kozakov, G.G. Levchenko, Liquid-phase sintered bismuth ferrite multiferroics and their giant dielectric constant, Ceramics International, 45 (2019) 14873–14879.

[8] A.V. Pashchenko, N.A. Liedienov, Q. Li, D.D. Tatarchuk, V.A. Turchenko, I.I. Makoed, V.Ya. Sycheva, A.V. Voznyak, V.P. Kladko, A.I. Gudimenko, Y.V. Didenko, A.T. Kozakov, G.G. Levchenko, Structure, non-stoichiometry, valence of ions, dielectric and magnetic properties of single-phase $Bi_{0.9}La_{0.1}FeO_{3-\delta}$ multiferroics, Journal of Magnetism and Magnetic Materials, 483 (2019) 100–113.

[9] A.V. Pashchenko, N.A. Liedienov, Li Quanjun, I.I. Makoed, D.D. Tatarchuk, Y.V. Didenko, A.I. Gudimenko, V.P. Kladko, Lina Jiang, Liping Li, V.G. Pogrebnyak, G.G. Levchenko, Control of dielectric properties in bismuth ferrite multiferroic by compacting pressure, Materials Chemistry and Physics, 258 (2021) 123925.

IOP Conf. Series: Materials Science and Engineering 1150 (2021) 012005 doi:10.1088/1757-899X/1150/1/012005

Laccase nanoparticle: Synthesis, characterization, entrapment in alginate beads and application in the biodegradation of Bisphenol A

Ademakinwa Adedeji N[1], Abidemi Ojo[2], Mariana Erasmus[2], Agunbiade Mayowa O.[3*]

[1]Department of Physical and Chemical Sciences, Elizade University, Nigeria
[2]Centre for Mineral Biogeochemistry, University of the Free State, Bloemfontein, 9301. South Africa
[3]Applied Microbiology and Health Biotechnology Institute, Cape Peninsula University of Technology, Cape Town, South Africa

E-mail: Agunbiadem@cput.ac.za

Abstract. Laccase nanoparticles (*LNP*) were synthesized by desolvation of the enzyme with ethanol and cross-linking with glutaraldehyde. *LNP* was characterized via scanning electron microscopy (*SEM*) and then entrapped in alginate beads for the biodegradation of bisphenol A (*BPA*). The stability of the immobilized *LNP* (*iLNP*) to heat inactivation was also investigated. The *iLNP* retained about 53±4.7% of its initial activity after 7 cycles of catalysis. The *iLNP* was stable to thermal inactivation at 40-70°C as obtained from its kinetic and thermodynamic parameters. The *iLNP* removed 80 mg/L *BPA* from solutions with a biodegradation efficiency (*BE*) of 92% after 1 h. Repeated use of the *iLNP* in *BPA* removal resulted in a BE of 75% after 7 cycles of catalysis. *iLNP* serves as a novel biocatalyst in the bioremediation of pollutants such as *BPA,* and its thermostable characteristics make it useful for other biotechnological applications

1. Introduction

Bisphenol A is toxic due to its estrogenic effects on the female reproductive systems, thus leading to BPA being termed as an endocrine-disrupting chemical [1]. The adverse effects of prolonged exposure to BPA include breast cancers, reduction in male sperm counts, immune disruption, to name a few. The effects on aquatic life are also deleterious as the fertility in fish is reduced [2]. Several techniques have been suggested to remove phenols and their derivatives; these include chemical oxidation, solvent extraction and even electrochemical methods. These techniques are often deficient as they lead to the release of more toxic by-products to the environment, they are too costly, and they are not very efficient [3]. The use of the biological approach is more cost-effective as it is efficient and generates non-toxic by-products. Examples of biological approaches include using enzymes produced by microorganisms and plants in the biodegradation of these phenols [4]. Enzymes that play key roles in the biodegradation of these phenols are laccases. Laccases (benzenediol: oxygen oxidoreductase EC 1.10.3.2) are oxidoreductases that play a role in the oxidation of a wide range of aromatics. Laccases have been employed in both free and immobilized form in the biodegradation of phenols and other pollutants such as polycyclic aromatic hydrocarbons, chlorophenols etc. A good review on how laccases are deployed in the biodegradation of pollutants such as BPA was provided by Bilal et al. [1]. Laccase for the biodegradation of phenols and derivatives are mostly obtained from the *Trametes* genus, and there is the need to explore other fungi capable of expressing these enzymes [5, 6]. As proposed in this study, we and other authors have noted that *Aureobasidium pullulans* are good producers of extracellular laccase [7, 8]. In reports where laccases (in free or immobilized forms) were used in the biodegradation of phenols, often, numerous setbacks are observed due to the relative poor catalysis and reduced reusability of the enzymes. Hence, there is the need for a novel and intuitive approach to immobilization of laccase for the biodegradation of these phenols. Therefore, we proposed using laccase nanoparticles entrapped in alginate beads. Enzyme nanoparticles are known to be more catalytically active, thermostable than free enzyme. In fact, enzyme nanoparticles are often used as biosensors because of their increased electrochemical properties [9]. This study is the first of its kind to explore the potential use of enzyme nanoparticles in the biodegradation of phenols.

2. Methodology

2.1. Microorganism and conditions

Aureobasidium pullulans was maintained on malt extract agar at 4°C. Sub-culturing was done every 72 h. Production of mycelia for extraction of intracellular laccase was done as described by Ademakinwa et al. [8]. The culture supernatant was centrifuged at 4500 x g for 15 min, and the mycelia obtained served as the source of intracellular laccase. Mycelia homogenization for extraction of the intracellular enzyme was carried out using techniques described by Ademakinwa and Agboola [10]

2.2. Extraction of intracellular laccase and assay

Mycelia obtained in section 2.1 above was homogenized in an equal volume of very chilled 50 mM phosphate buffer pH 6.5. The homogenate was centrifuged at 13500 x g for 20 min to obtain a clear supernatant that serves as the intracellular laccase. Laccase was assayed for using ABTS as substrates respectively [7]

2.3. Preparation, characterization and immobilization of laccase nanoparticles

Laccase nanoparticles (LNP) were prepared according to the methods described by Jahkar and Singh et al. [9]. Briefly, laccase was desolvated using ethanol and cross-linked using glutaraldehyde. The laccase nanoparticles were purified by centrifugation at 10000 x g for 10 min. The LNP was stored at 4°C for further use. The LNP was characterized using scanning electron microscopy [9]. Immobilization of the nanoparticles on the alginate complex was carried out as described by the methods of Ademakinwa and Agboola [12] and Bezbaurah et al. (13) with minor modifications.

2.4. Thermodynamic and kinetic properties of laccase nanoparticles entrapped in alginate support

The thermal inactivation process was determined according to the methods described by Ademakinwa et al. [14], Ademakinwa and Agboola [10] and Ademakinwa et al. [12] after the determination of the temperature profile.

2.5. Biodegradation of phenols and bisphenol A in model wastewater

The biodegradation of bisphenol A was performed according to the methods of Asagdol et al. [11]. The reaction mixture contained 0.1 g of immobilized LNP and phenol or BPA solution (final concentration of 80 mg/L) followed by incubation at 35°C and stirring at 50 rpm for 40 min. Samples were withdrawn every 10 min and analyzed for residual concentrations of phenolic pollutants.

3. Results and Discussion

Fig.1 shows the SEM images for the laccase nanoparticles. The immobilized laccase nanoparticles (iLNP) were retained in the alginate beads, and they oxidized ABTS. The iLNP were used for several rounds of ABTS catalysis and the iLNP retained about 53% of residual activity after 7 catalytic cycles (Fig. 2). A gradual loss in enzyme activity might be associated with enzyme leakage from the beads [16]. Table 1 summarizes the kinetic and thermodynamic properties of the free laccase and iLNP. The heat denaturation process generally followed first-order reaction kinetics. This observation has been noted in

Figure1. Scanning electron microscopy images of Laccase Nanoparticles (A) x100 (B) x 500

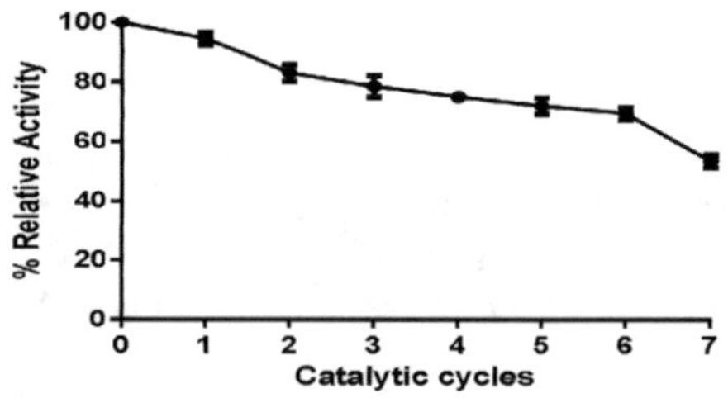

Figure 2. *iLNP* catalysis of ABTS oxidation. The experiment was performed in triplicate, and error bars indicate deviation from mean

enzymes produced by this fungus [10, 12, 14, 15]. It can be observed that generally, the half-life ($t_{1/2}$) values decreased as the temperature was increased from 50 to 80ºC.

Table 1. The kinetic and thermodynamic parameters for free laccase and *LNP* thermal inactivation process

***T* (*K*)**	**$k_d\ (h^{-1})$**	**$t_{1/2}$ (*h*)**	**ΔH^*_d (kJ/mol)**	**ΔG^*_d (kJ/mol)**	**ΔS^*_d J/mol/K**
40	#0.031	#22.0	#37.8	#88.3	#-161.7
	*0.010	*69.0	*42.4	*89.9	*-151.2
50	#0.070	#11.0	#37.5	#88.8	#-161.3
	*0.011	*63.0	*42.3	*92.2	*-154.4
60	#0.170	#4.0	#37.3	#86.9	#-156.8
	*0.019	*37.0	*42.2	*93.4	*-153.4
70	#0.180	#4.0	#37.1	#89.8	#-155.2
	*0.0200	*35.0	*42.1	*94.9	*-152.5
80	#0.200	#4.0	#37.1	#90.3	#-155.1
	*0.030	*23.0	*42.1	*95.3	*-151.0

#- Free laccase

*- Immobilized Laccase Nanoparticles (*iLNP*)

The result showed that the denaturation of the iLNP and the free enzyme increased as the temperature increased. Meanwhile, k_d, which is the inactivation constant decreased as the temperature increased both in the free laccase and iLNP. Comparatively, the values obtained for the kinetic parameters indicated that the iLNP was thermally stable to thermal inactivation than the free enzyme. From the values obtained for thermodynamic parameters (for the free laccase and laccase iLNP) such as $\Delta H_d^*, \Delta S_d^*$ *and* ΔG_d^* , it can be concluded that the iLNP showed better resistance to heat inactivation and hence it could be useful for other biotechnological applications. For example, the entropic values were negative; this implies that there was resistance to disorderliness as the temperature increased. And overall, this indicated thermostability to thermal denaturation. The iLNP was able to efficiently biodegrade BPA with a biodegradation efficiency of 76% after 2 h. The biodegradation efficiency of the iLNP decreased to 56% after the third catalytic cycle. The decrease might be associated with leakage of the LNP from the beads [16].

4. Conclusion

Laccase nanoparticles synthesized via de-solvation with ethanol, cross-linking with glutaraldehyde and immobilized by entrapment in alginate displayed unique properties such as resistance to thermal inactivation. The LNP was able to efficiently biodegrade BPA resulting in a biodegradation efficiency of 76%. The heat resistant properties of the iLNP suggest that it can be a novel biocatalyst for numerous industrial applications.

References

[1] Bilal M, Iqbal HMN, Barceló D.2019. Mitigation of bisphenol A using an array of laccase-based robust bio-catalytic cues – A review. Science of The Total Environment. doi:10.1016/ j.scitotenv. 2019.06.403
[2] Chhaya U, Gupte A. 2013. Possible role of laccase from Fusarium incarnatum UC-14 in bioremediation of Bisphenol A using reverse micelles system. Journal of Hazardous Material, 254–255:149–156.
[3] Modaressi K, Taylor KE, Bewtra JK, Biswas N. 2005. Laccase-catalyzed removal of bisphenol-A from water: protective effect of PEG on enzyme activity Water Research. 39:4309–4316.
[4] Inoue M, Masuda Y, Okada F, Sakurai A, Takahashi I, Sakakibara. M 2008. Degradation of bisphenol A using sonochemical reactions. Water Research. 42:1379–1386.
[5] Roy S, Ishita D, Munjal M, Karthic L, Kumar G, Kumar S, Rao KVB. 2014. Isolation and characterization of tyrosinase produced by marine actinobacteria and its application in the removal of phenol from the environment. Frontiers in Biology 9(4):301-316.
[6] Ryan D, Leukes W, Burton S. 2007. Improving the bioremediation of phenolic wastewaters by Trametes versicolor. Bioresource Technology, 98:579–587.
[7] Rich JO, Leathers TD, Anderson AM, Bischoff KM and Manitchotpisit P. 2013. Laccases from Aureobasidium pullulans. Enzyme and Microbial Technology, 53(1), 33–37. doi:10.1016/j.enzmictec. 2013.03.015
[8] Ademakinwa AN, Agboola FK. 2016. Biochemical Characterization and Kinetics of a Purified Yellow Laccase from Aureobasidium pullulans (De Bary) Isolated from Soil Containing Decayed Plant Matter. Journal of Genetic Engineering and Biotechnology 14(1):143-151.
[9] Jakhar S, Pundir S. 2018. Preparation, characterization and application of urease nanoparticles for construction of an improved potentiometric biosensor. Biosensor and Bioelectronics 100: 242-250

[10] Ademakinwa AN, Agboola FK. 2020. Some biochemical, catalytic, thermodynamic and kinetic properties of purified fructosyltransferase from wild and improved mutant-type Aureobasidium pullulans NAC8. Biocatalysis and Biotransformation, 1–12. doi:10.1080/10242422.2019.1671376
[11] Asagdol Z, Forootanfar H, Rezaei S, Mahvi AH, Faramarzi MA. (2013). Removal of phenol and bisphenol-A catalysed by laccase in aqueous solution. Journal of Environmental Health and Engineering 12:93 (1-5)
[12] Ademakinwa AN, Agboola FK. (2019). Kinetic and Thermodynamic Investigations of Cell-Wall Degrading Enzymes Produced by Aureobasidium pullulans via Induction with Orange Peels: Application in Lycopene Extraction. Preparative Biochemistry and Biotechnology. doi: 10.1080/10826068.2019.1650375
[13] Bezbaruah, A. N., Krajangpan, S., Chisholm, B. J., Khan, E. Elorza Bermudez, J. J. (2009). Entrapment of iron nanoparticles in calcium alginate beads for groundwater remediation applications. Journal of Hazardous Materials, 166(2-3), 1339–1343. doi:10.1016/j.jhazmat.2008.12.054
[14] Ademakinwa AN, Agunbiade MO and Fagbohun O. 2020. Biodegradation of cyanide in cassava wastewater using a novel thermodynamically-stable immobilized rhodanese. Preparative Biochemistry & Biotechnology, 1–11. doi:10.1080/10826068.2020.1846053
[15] Ademakinwa AN, Agunbiade MO, Ayinla ZA, Agboola FK (2019). Optimization of Aqueous Two-Phase Partitioning of Aureobasidium pullulans α-Amylase via Response Surface Methodology and Investigation of its Thermodynamic and Kinetic Properties. International Journal of Biological Macromolecules doi: https://doi.org/10.1016/j.ijbiomac.2019.08.159
[16] Lassouane F, Aït-Amar H, Amrani S, Rodriguez-Couto A. 2019. A promising laccase immobilization approach for Bisphenol A removal from aqueous solutions. Bioresour. Technol. doi:10.1016/j.biortech.2018.09.129

Study on the effect of heat exchange tube structure on waste heat recovery capacity of the internal combustion engine in the freshwater distillation system

Dien Vu Minh[1,2], Quang Khong Vu[1,*], Tien Nguyen Duy[1], Le Manh Toi[1]

[1]Department of Internal Combustion Engine, School of Transportation Engineering, Hanoi University of Science and Technology, 1 Dai Co Viet Road, Hanoi, Viet Nam
[2]Center for Automotive Technology, Hanoi University of Industry, 298 Cau Dien Road, Hanoi, Viet Nam

E-mail: quang.khongvu@hust.edu.vn

Abstract. In the Internal Combustion Engine (ICE) working operation, fuel is fed into the combustion chamber, where it is burnt in air to convert the chemical energy of fuel into heat according to the first law of thermodynamics. However, the amount of energy transformed into actual power is only 21-33% and 25-40% for gasoline engines and diesel engines, respectively. Consequently, about two-thirds of the remaining energy is lost due to the heat transfer to the surrounding environment, in which the heat energy removed by the exhaust occupies the largest proportion (about 30-35%). Therefore, utilizing the regenerative heat from the exhaust gases is a potential solution to improve the ICE's efficiency. This paper presents the simulation results of the effect of heat exchange tube structure on waste recovery capacity in the system, which utilizes the heat of exhaust gases and cooling water of ICE based on Ansys Fluent. In addition, the boundary conditions, the flow, and exhaust temperature are determined by simulating the engine's working process using AVL-Boost. This research shows that the heat utilization efficiency of exhaust energy can be achieved up to 34%, corresponding to a reasonable structure. This research's study outcomes are the foundation to implement the complete design of the heat waste recovery system of ICE.

1. Introduction

Increased thermal efficiency is always an essential goal in improvement research of Internal Combustion Engine (ICE). According to applied modern technologies, the thermal efficiency of current ICE is only 30-40%. As a result, there is still a large amount of heat energy emitted into the environment, corresponding to 60-70% of total energy [1-2]. Moreover, due to the pressure on reserves of fossil fuel sources, the price is fluctuating and has an increasing tendency; energy security contains many latent risks, while the requirement for ICE emissions has been tightened [3-5]. As a consequence, the current automobile manufacturers have to compete to increase heat efficiency and reduce toxic emission. Therefore, many eco-friendly technologies have been deployed in worldwide applications. One of the potential solutions to improve thermal efficiency, which is being studied by many researchers in this field, is recovering the heat energy from exhaust gases and cooling water of ICE. For example, M. Hatami et al. performed some research to optimize the ICE exhaust gas utilizer tube's structure to produce actual power [6]. E.H Wang et al. revealed

IOP Conf. Series: Materials Science and Engineering 1150 (2021) 012006 doi:10.1088/1757-899X/1150/1/012006

and specified some organic liquid for the Rankine cycle to improve the efficiency of utilizing heat from ICE exhaust gases [7].

However, these solutions still not exploit the most energy of the waste heat. The previously calculated results are only exact at a stable engine regime with a large number of exhaust gases. Meanwhile, in real situations, ICE always works at variable loads and speeds, leading to the change of the amount of waste heat, which affects the heat utilizing the efficiency of the system. Thus, in order to maximize the efficiency and the basis to design the systems that can achieve higher efficiency in different engine regimes. This paper will focus on researching and evaluating the influence of structural parameters on the ability to utilize the heat of exhaust gases of exchange tube by Ansys-Fluent. The thermodynamic parameters, boundary conditions of the model, were determined through the simulation process on AVL-Boost in cases of D243 engine operating at 40%, 60%, 80%, 100% loads, speed 2200 rpm.

2. Determine the boundary conditions of the model

2.1. Building the engine model

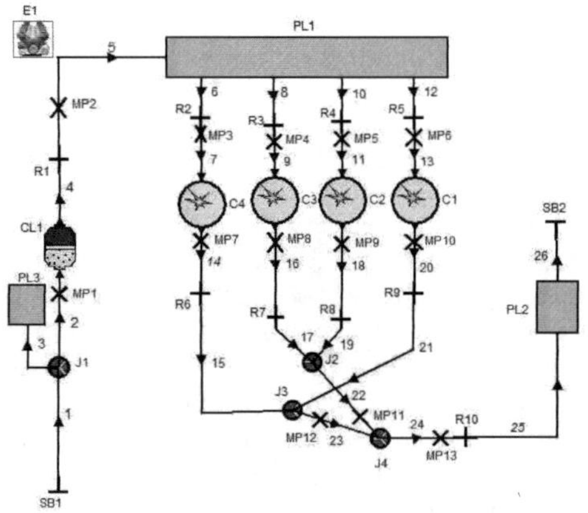

Figure 1. Simulation model of D243 engine in AVL-Boost

AVL-Boost is a one-dimension simulation software based on the combination of the first law of thermodynamics with the calculation models of combustion, heat transfer, and gas exchange [11]. Based on the D243 engine structure, measurement parameters and related documents, the elements of the simulation model are selected appropriately and shown in Fig 1. The reliability of the model is evaluated through deviations of N_e and g_e of the engine between simulations and experiments. To specify these parameters, the D243 engine was tested and measured on the engine testbed at the research center for engines, as shown in Fig.2.

The testbed consists of the following main devices: AVL APA 100 dynamometer, Oil cooler AVL 554, Coolant system AVL 553, AVL fuel balance 733S, Fuel temperature control AVL 753, Throttle actuator THA 100 and PUMA software (as shown in Fig 2a).

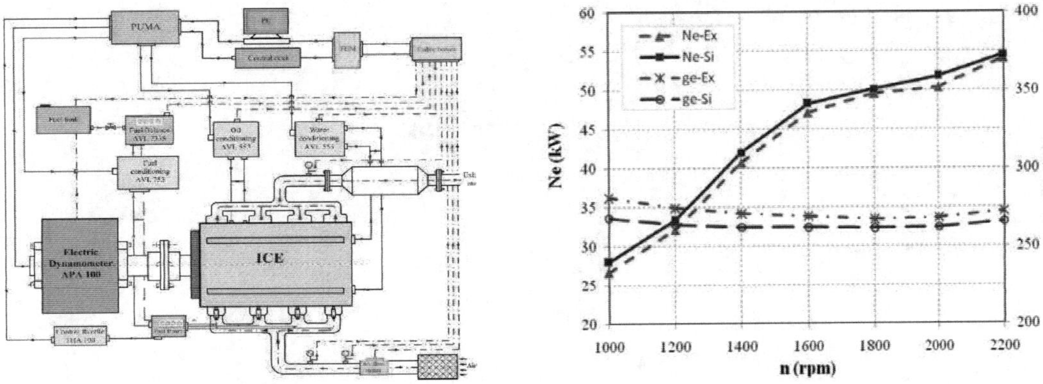

a) b)

Figure 2. Schematic arrangement of engine testbed and comparison results between simulation and experiment

After inserting the data into the simulation model, the process of running and modifying the model was completed based on the experiment results. This process only stops when the simulation results narrow the gaps with the experiment. Then continuously running the model at the points of maximum fuel mass injection to compare between the simulation (Si) and the experiment (Ex) results. The reliability of the model is evaluated by the difference between the Si and Ex's power and fuel consumption, as shown in Fig.2b. The results indicate that the values of Ne and ge between Si and Ex is very similar. All of the deviations are less than 5%, in which the greatest error of N_e is 4.87% and g_e is 4.65% at n = 1000 rpm.

2.2. The simulation results determine boundary conditions for heat exchanger

With the model of engine achieved the necessary reliability, the next step is running the simulation to determine the thermodynamic parameters of exhaust gases, such as temperature (T_{exh-in}), flow rate (V_{exh}) and heat rate (Q_{exh}) of exhaust gases according to loads at speed of 2200 rpm; the results are shown in Table 2.

Table 2. The thermodynamic parameters of exhaust gases at 2200 rpm

% Load	T_{exh-in} (K)	V_{exh} (m/s)	Q_{exh} (kJ/s)	q_{exh} (%)
40	704.4	85.3	39.49	36.3
60	761.7	91.2	45.80	35.0
80	815.7	96.6	51.87	33.9
100	870.7	101.9	58.20	33.2

2.3. Proposed model for utilizing exhaust heat

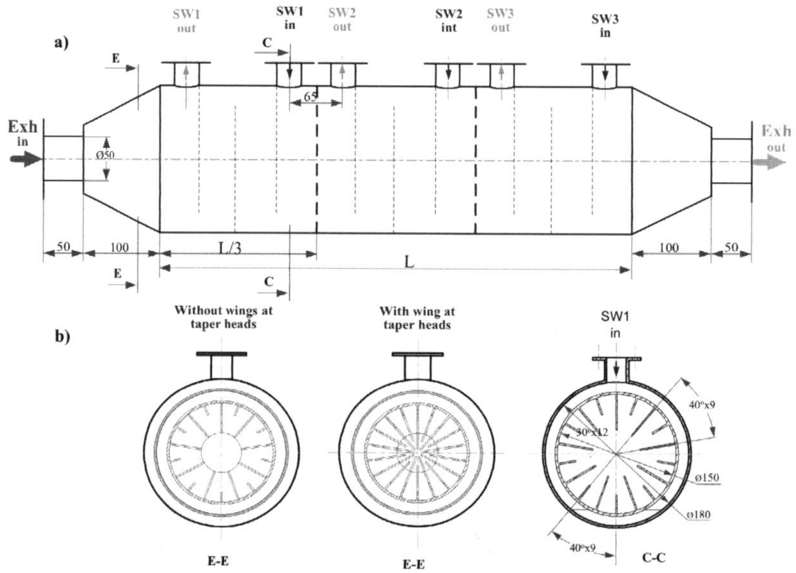

Figure 3. Design of heat exchange tube

Based on the thermodynamic parameters of exhaust gases (Table 2), the design of the heat exchange tube is described in Fig.3. The tube has a cylindrical structure similar to the sound absorption. The exhaust gases normally come into the center of the tube and exchange heat with the wings inside the heat exchanger (18 wings with 9 long and 9 shot, the wings are arranged alternately); the outside is divided into three seawater compartments along the length of the tube. In addition, there are wing directions in each water compartment to increase the ability to exchange heat between water and exhaust gases. Regarding to such structure, the heat exchanger is installed in the exhaust system of the test engine with the exhaust gases moving along inside and the seawater surrounding of the tube, seawater, and exhaust gases are on opposite directions.

3. Results and Discussion

3.1. The effect of structure on the ability to utilize exhaust energy

With the tube length of 600 mm, the impact of structure on the heat exchange process between the exhaust gases and seawater are shown in Figs.4-5. In Fig.4, the flow of exhaust gases moving inside the tube has a temperature changing in different regions. In two cases, the temperatures of the exhaust gases are generally reduced from the input to the output, at the positions closed to the tube wall or the heat exchange wings, the exhaust gases temperature is reduced compared to the other positions. This phenomenon can be explained that in areas with lower temperatures, there is heat exchanged between exhaust and seawater.

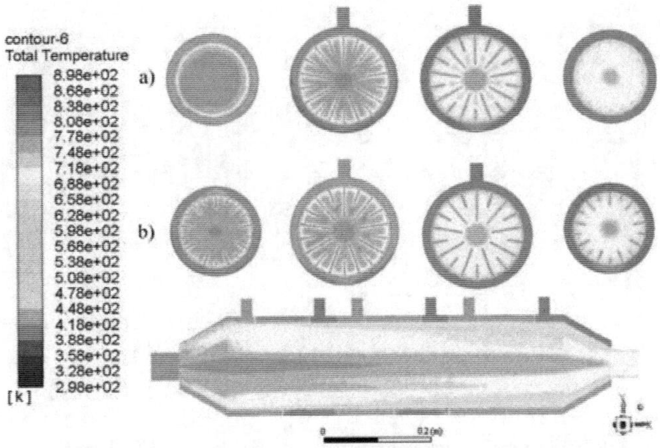

Figure 4. Temperature distributions inside the heat exchanger without and with wings at two taper heads in 100% load and 2200 rpm. a) Without wings and b) With wings

Fig 5 shows the effect of the arrangement of wings on the heat exchanger, the recovery efficiency, and the seawater flow through the tube. The results indicate that when adding wings on the taper heads, the utilization efficiency, as well as the flow of seawater through the tube, is better. Specifically, the maximum value of the heat exchange, the recovery efficiency, and the total seawater flow reach 34.11% and 7.88 lit/min at 100% load of the engine, respectively.

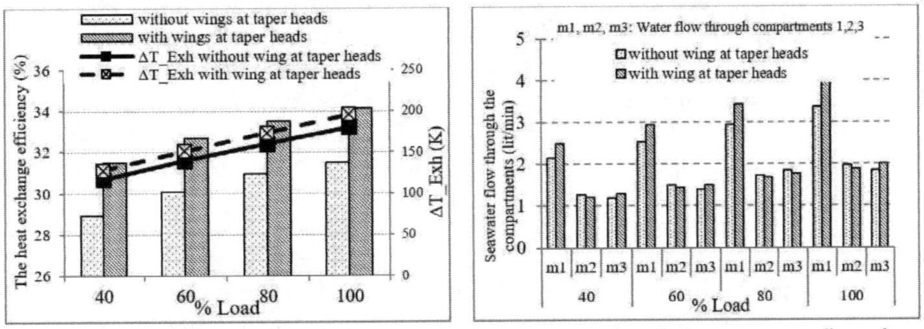

Figure 5. Effect of structure, with and without wings at taper heads on seawater flow through the compartments, on the heat exchange efficiency, and the decrease of exhaust gases temperature

3.2. The effect of length pipe on the ability to utilize exhaust energy

The increase of heat exchange area by the arrangement of wings on both of the taper heads significantly increases the heat exchange capacity of the tube. However, this option can redouble the mobility of exhaust gases. Therefore, this research performs further investigation of the ability to exchange heat through the tube by changing the length at three levels of 600, 645, and 690 mm in the case without wings at taper heads.

The relationship between the length of tubes and the heat recovery capacity of the system is shown in Fig.7. The result shows that when the length of the tube (L) changes from 600 mm to 690 mm, the recovery efficiency (η) increases from 31.48% to 34.12% and the total seawater flow (lit/min) through the compartments increases from 7.18 lit/min to 7.79 lit/min at 100% load and 2200 rpm.

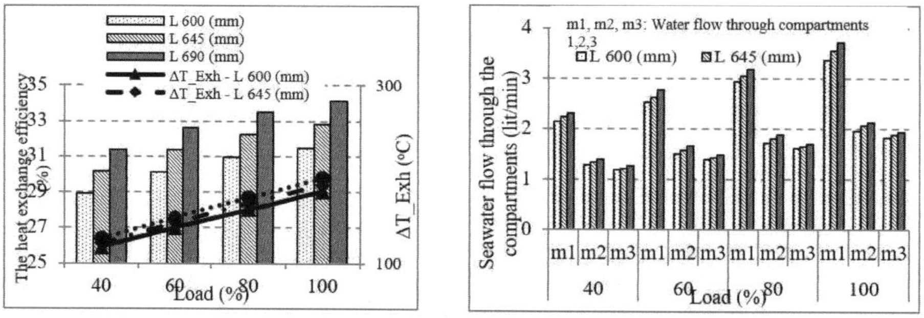

Fig. 6. Effect of length on seawater flow through the compartments, on the heat exchange efficiency and the decrease of exhaust gases temperature

The above results can be explained when the pipe length increases, the heat exchange area between waste gas and seawater is improved and thus facilitating the heat exchange process between exhaust and seawater. In addition, when increasing the load, the exhaust energy increases both in heat and kinetic energy, it also increases the heat exchange capacity between the exhaust gases and seawater in the heat exchange tube.

4. Conclusions

This paper was successful in building heat exchange tube models as well as evaluating the effect of structural parameters on heat exchangeability. The simulation results show that to increase the heat exchange capacity between exhaust gases and seawater, it is necessary to increase the area of heat exchange by adding wings or increasing the tube's length. However, it should be noted that when adding wings, attention should be paid to the ability to make as well as obstructing mobility of exhaust gases. In the meantime, increasing the length of tubes will affect the layout space of the heat exchange tube in the system. This study's results are going to be an important factor in calculating, designing, and implementing the system utilizing the useless energy transferred to exhaust gases of the ICE.

Acknowledgment:

This work was supported by the Ministry of Education and Training for the financial support under the grant contracts No.2017 BKA 39. This research was also assisted by research center for engines, fuels and emissions at Hanoi University of Science and Technology for the test equipment as well as the AVL-Boost program

References

[1] J. S. Jadhao and D. G. Thombare, "Review on Exhaust Gas Heat Recovery for I.C. Engine," International Journal of Engineering and Innovative Technology, Volume 2, Issue 12 (June 2013), pp 93-100.

[2] Quang Khong Vu, Dien Vu Minh, Tien Nguyen Duy, Tuan Pham Minh and Luong Nguyen The, "A Study of Exhaust Waste Heat Recovery in Internal Combustion Engines," IOP Conf. Series: Materials Science and Engineering 507 (2019) 012028.

[3] Giorgio Zamboni, Simone Moggia and Massimo Capobianco, "Hybrid EGR and turbocharging systems control for low NOX and fuel consumption in an automotive diesel engine," Applied Energy 165 (2016), pp 839–848.

[4] Kai Shen, Fangbo Li, Zhendong Zhang, Yuedong Sun and Congbo Yin, "Effects of LP and HP cooled EGR on performance and emissions in turbocharged GDI engine," Applied Thermal Engineering 125 (2017), pp 746–755.

[5] Jiaqiang Ea, Dandan Hana, Yuanwang Denga, Wei Zuoa, Cheng Qian, Gang Wu, Qingguo Peng and Zhiqing Zhang, "Performance enhancement of a baffle-cut heat exchanger of exhaust gas recirculation," Applied Thermal Engineering 134 (2018), pp 86–94.
[6] M. Hatami, D.D.Ganji and M.Gorji-Bandpy, "Numerical study of finned type heat exchangers for ICEs exhaust waste heat recovery," Case Studies in Thermal Engineering 4 (2014), pp 53–64.
[7] E.H. Wang, H.G. Zhang, B.Y. Fan, M.G. Ouyang, Y. Zhao and Q.H. Mu, "Study of working fluid selection of organic Rankine cycle (ORC) for engine waste heat recovery," Energy 36 (2011), pp 3406-3418.

IOP Conf. Series: Materials Science and Engineering 1150 (2021) 012007 doi:10.1088/1757-899X/1150/1/012007

Biomimetic approaches to color formation in ecological architecture

O Krivenko[1], O Pylypchuk[2], N Avdieieva[3], M Avdieieva[4] and S Bieber[5]

[1] Department of Architectural Constructions, Kyiv National University of Construction and Architecture, 31 Povitroflotsky Avenue, Kyiv, 03037, Ukraine
[2] Department of Drawing and Painting, Kyiv National University of Construction and Architecture, 31 Povitroflotsky Avenue, Kyiv, 03037, Ukraine
[3] Department of Architectural Design Deal, SSU «Institute of Innovative Education of Kyiv National University of Construction and Architecture»,4 Education Street, Kyiv, 03037, Ukraine
[4] Department of Foundation of Architecture, Design and Urban Planning, National Aviation University, 1 Lubomir Husar Avenue, Kyiv, 03058, Ukraine
[5] Department of Architectural Design Deal, SSU «Institute of Innovative Education of Kyiv National University of Construction and Architecture»,4 Education Street, Kyiv, 03037, Ukraine

E-mail: pylypchuk.od@knuba.edu.ua

Abstract. The article discusses the characteristics and features of the color formation in nature for solving problems of ecological architecture. The analysis of the main directions of research on the significance and possibilities of color solutions in creating an ecologically sustainable urban environment is presented. It is proposed to use analogs of nature to obtain natural features and qualities of ecological design of buildings, including in the formation of the color structural component of the design. As a result of the analysis of the studies, the mechanisms of color formation in nature, based on functional expediency, adaptation to the conditions of life in the environment, have been identified. The mechanism of formation of pigment, structural and combined colors in nature is considered. On the basis of the presented analysis of the functionality of color formation in nature, coloristic means of regulating the energy-efficient and environmental parameters of the building, namely thermoregulation, regulation of illumination, humidity, are identified. The examples of color solutions of building finishing aimed at adaptation to external environmental conditions are analyzed. The methods and possibilities of forming the color structure on the basis of the studied natural analogues for solving the problems of modern ecological architecture are systematized.

1. Introduction

After a period of dominance of purely technical solutions at the present stage of architecture development, more and more attention is paid to ecological design, the purpose of which is to create healthy and comfortable living conditions for a person while reducing energy costs and environmental impacts. The predominantly artificial objective environment of a modern city requires a return to design compatible with the environment, climate, that is, ecological design.

In total, in modern design, actual original spatial-shaped concepts are associated with new trends and stylistic directions, and, most importantly, with the development of new technologies. The reference to biological examples that inspire engineers to create new materials and technologies is based on the assumption that over billions of years of evolution, nature has created optimal living structures that are superior in efficiency and durability to man-made structures [1, 2].
Biomimetics is a design direction based on the reproduction of ideas from nature into constructive and technological solutions [3, 4].

The development of biomimetic approaches is a promising direction for increasing durability and efficiency in construction, creating new directions for the development of modern ecological architecture aimed at adaptation to the environment.

The study proposes to consider the directions and prospects for the development of biomimetic approaches in creating color solutions for modern ecological architecture.

2. Formation of biomimetic color formation approach

The development of various effective approaches to the design of ecological architecture requires a developed complex apparatus of various scientific knowledge, as well as the formation of a base of analogues of the natural world, biological systems, including the inclusion of the coloristic component of education.

2.1. Color application in architecture

From the point of view of the means of an architect and designer in creating a subject space, color is one of the main tools in the formation of compositional solutions in accordance with the selected color range and the appropriate use of their optical properties. This is achieved due to the texture and texturelessness of the colored surface (texture of the material, contrast, visual memory, special observation conditions) [5]. Revealing or masking the texture and texture of a material using color is closely related to tectonics – the artistic expression of essential physical and decorative properties through the plasticity of the form [6]. The correspondence between material and form is an important condition for overall tectonicity, aesthetics and harmony. According to research, an important role is played by the material, its physical and technical capabilities, as well as the joint work of material and structure [7]. The formation of the visual perception of buildings in coloristic design is influenced by: a) natural and climatic features; b) coloration of the urban environment; c) associative and psychological impact of color d) color-compositional solutions, taking into account tectonics, shape, scale, texture; e) traditional perception of color, taking into account the socio-cultural characteristics of the area.

2.2. Color formation in nature

The analysis of the research made it possible to determine that the harmony of coloristic solutions in nature is associated with the provision of tasks of functional expediency, optimization of adaptation to the environment and living conditions due to the protective or camouflage function; communication signal; thermoregulation; optimization of the spectrum of rays absorbed by leaves during photosynthesis [1, 8, 9].

In living and inanimate nature, color change is a reaction to changing environmental conditions. Coloring in nature is characterized by various kinds of variability: a) irreversibility with time or age, for example, how leaves change color according to seasons, which is associated with the synthesis, accumulation and decay of pigments [8]; b) reverse phenomena, when a color change occurs as a reaction to changing environmental factors and is provided by the movement of pigment granules or special pigment cells (melanophores). This color reaction is associated with the regulation of temperature and humidity: for example, the body of a lizard becomes darker with a decrease in temperature, so that dark colors absorb more heat, and conversely, with an increase in temperature, a lizard can become very pale because light colors reflect heat; Hercules beetle (Dynastes hercules) has a greenish-red color at normal humidity, however, if the air humidity exceeds 80%, the color changes to black due to the filling of air cavities with moisture in the structure of the chitinous cover of the elytra [8].

The process of evolution and adaptation of color in nature optimizes the mechanism of its formation. As a result of the analysis of research, the following mechanisms of color formation in nature should be distinguished:

1. Pigment (chemical), which is associated with the ability of certain molecules (pigments) to selectively absorb, reflect or emit light with a certain wavelength. Pigmentation in living nature occurs by its own synthesis or through the process of nutrition [8].

2. Structural (other names – iridescence or irisation), where the formation of color is based on physical processes and depends on the microstructure of surfaces, on which light from the source falls, causes diffraction or interference of light [9]. In all cases, irisation is based on nanostructures in the form of ribs, fibers, plates, organized in regularly spaced rows or lattices (in physics, structures of this type are called photonic crystals). Photonic crystals create specific optical effects such as diffraction and interference. For the interference effect to occur, it is necessary that the light waves repeatedly reflected from the array elements are in the same phase. The amplitudes of the waves for which this condition is met are added, and the wavelengths of these waves determine the color background [9].

3. Combined – a combination of structural and pigment mechanisms, enhances their action and forms a variegated palette [9].

Thus, the variety of colors in nature is created due to evolution, is built according to the laws of optical physics and a complex system of the structure of living cells and tissues, has its own content and function. The process of evolution and adaptation of color in nature optimizes the mechanism of its formation.

3. Examples of the formation of a biomimetic approach to the use of color in solving problems of ecological architecture

The use of natural analogues of color formation in ecological architecture makes it possible to apply new, more energy-saving, natural, ecological and durable mechanisms for obtaining color solutions for home design. For example, each color has certain strength of absorption and reflection of light, depending on its spectral composition and light sources. If the color belongs to a certain texture of the material, then the effect of its reflection, or the strength of brightness will change, and accordingly the perception of the form, the carrier of the given color, will also change. Understanding the functional basis of the formation of colors in nature makes it possible to consciously use color as a means of regulating energy efficient and environmental parameters in ecological architecture (table 1).

Table 1. Application of natural analogues of color formation in ecological architecture.

№	Functions	Natural solutions	Application in ecological architecture
1	Thermore-gulation	Natural mechanisms for the formation of visible color and gloss, affecting the degree of heat absorption	Thermoregulation through the color of the façade (light to dark, matt to glossy) will provide varying degrees of heat absorption
2	Humidity control	Natural mechanism of color return when external climatic parameters change	The use of a "smart facade" in the system, which can react to the degree of moisture or heat absorption by changing color (for example, nanofilm – moisture indicator, changes color [10])
3	Ensuring durability	Structural mechanism of color formation	Application of the natural mechanism of the structural formation of color in the decoration of facades

4	Light control	Natural mechanisms for the formation of visible color and gloss that affect the light reflection degree	Regulation of illumination of facades of buildings, streets, premises when using different colors, is provided by different degrees of light reflection for different colors

In connection with the development of the environmental direction in the design of buildings, today there are color solutions aimed at ensuring environmental compatibility with the environment. They are episodic and non-systematic, but reflect the trends in the development of a new ecological approach in the formation of color design of buildings, for example (fig. 1 (a-c)).

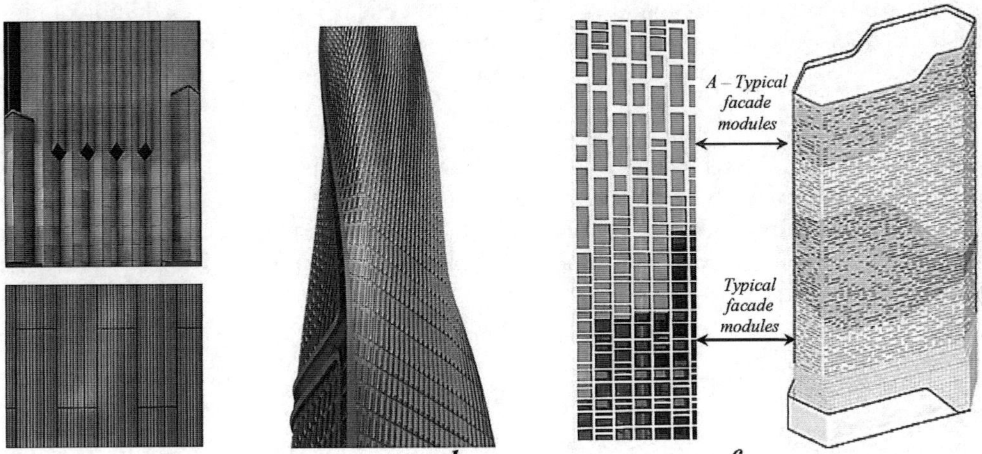

Figure 1 (a–c). Examples of using biomimetic approaches in shaping color in ecological architecture [11-13].

Key to Figure 1 (a-c):

(a) – Metropolitan Workshop's tower, a 27-storey building in London covered with turquoise glazed structured terracotta tiles, like a chameleon constantly changing color under different lighting conditions, reflecting the color of the river and plants around the house, and also has a natural self-cleaning function [11];

(b) – Wasl Tower, Dubai, United Arab Emirates, glazed clay facade tiles protect against thermal stress entering the building and the impact on its surroundings, providing an effective solution for thermal control of a high-rise building in a desert climate [12];

(c) – Seoul tower, in the Seoul, South Korea has the color scheme of the building facade determined by the geometry and dimensions of the facade elements, taking into account the influence of the Sun and orientation to the cardinal points to ensure user comfort inside the building and reduce energy consumption [13].

3.1. Discuss

In architecture, color is used as an important means of visual impact in an urban environment, in the formation of visual perception of the form, texture, material of a building. A significant number of studies are devoted to these issues [10, 14]. For the development of ecological architecture, it is promising to use analogs of coloristic solutions in nature, based on functional feasibility, optimization of adaptation to the environment, living conditions, are studied in various fields of studying nature [1, 2]. It should be noted that there are no systemic studies on the formation of approaches to the use of various means of color formation in living and inanimate nature to solve problems of ecological design and architecture, namely, to ensure their durability and to regulate energy-efficient and environmental parameters (thermoregulation, illumination regulation, humidity). The approach proposed in the article for the introduction of means and approaches for the formation of color in nature, as the basis for color education in accordance with functional and applied application, makes it possible to develop and solve the problems of ecological architecture.

4. Summary

Research results make it possible to establish that today the use of biomimetic approaches to the use of color in the design of buildings determines new trends in the creation of ecological solutions for facades based on the use of natural analogues. An analysis of the use of modern design innovative color solutions in buildings shows the relevance of the dissemination and development of new biomimetic ideas and their implementation in design and design practice in order to create an aesthetic, ecological and energy-efficient urban environment in accordance with environmental conditions. The addition of biomimetics, which defines approaches to nature as a source of knowledge and requirements for building design, provides the development of ecological architecture and a direction for further research.

References

[1] Dumanlia A G Ë and Savin T 2016 Recent advances in the biomimicry of structural colours *The Royal Society of Chemistry* **45** pp 6698–6724
[2] Krivenko O, Mileikovskyi V and Tkachenko T 2018 The principles of energy efficient microclimate provision in the skyscraper «Biotecton» of 1 km height European *J. of Formal Sciences and Engineering* **3** pp 8–17
[3] Lodson J and Jahromi F S 2017 Innovative Materials for Interior Architecture Sustainable S*tructure and Materials* **1** pp 1–11
[4] Gruber P and Imhof B 2017 Patterns of Growth – Biomimetics and Architectural Design *Buildings* **7** (32) pp 2–17
[5] Nemcsics A and Takacs J 2014 Experimental Determination of Laws of Color Harmony Part 8: Harmony Content Versus Relative Surface Coverage *Color Research & Application Wiley Period.* **39** pp 387–398
[6] Priyomarsono N W, Surya R and Budiman D S 2019 Tectonic in Architecture in Capability of Capturing Epoch The 1st International C*onference on Computer Science Engineering and Technology* (Tasikmalaya Indonesia 27–28 November 2018) IOP Publishing, Conf. Series: *J. of Phys.* **1179** p 012096
[7] Caivano J L and Giglio M P 2015 Aportes al estudio de la apariencia visual en contexto de prácticas proyectuales: arte, diseño y ambiente construido (Universidad Nacional de Mar del Plata) p 127
[8] Rassart M, Colomer J.-F, Tabarrant T and Vigneron J. P 2008 Diffractive hygrochromic effect in the cuticle of the hercules beetle Dynastes Hercules *New J. of Phys.* **10** 033014 pp 1–14
[9] Seo H-B and Lee S-Y 2017 Bio-inspired colorimetric film based on hygroscopic coloration of longhorn beetles (Tmesisternus isabellae) *Scientific reports* **44927** pp 1–8
[10] Zhao Y, Zhao Y, Hu S, Lv J, Ying Y, Gervinskas G and Si G 2017 Artificial Structural Color Piels: *A Review Plasmonics and Metamaterials* **10** (8): 944 pp 2–17
[11] Metropolitan Workshop covers tower of modular homes with glazed ceramic tiles *Dezeen Magazine* URL: https://www.dezeen.com/2019/01/15/mapleton-crescent metropolitan-workshop-tinyhomes-pocket/ (date retrieved: 17.01.2021)
[12] UNStudio Reveals Recent Construction Images of Wasl Tower in Dubai URL: https://www.archdaily.com/933286/unstudio-reveals-recent-construction-images-of wasltower-in-dubai (date retrieved: 17.01.2021)
[13] UNStudio's dynamic facade will control indoor climate for Seoul tower *Dezeen Magazine* URL: https://www.dezeen.com/2014/04/30/hanwha-hq-seoul-unstudio-dynamic-facade/ (date retrieved: 17.01.2021)
[14] Caivano J L 2006 Research on Color in Architecture and Environmental Design: Brief History, Current Developments, and Possible Future *Wiley COLOR* **8** pp 350–363

IOP Conf. Series: Materials Science and Engineering 1150 (2021) 012008 doi:10.1088/1757-899X/1150/1/012008

Study on the water vapor adsorption Characteristics of Carbon materials in HTGR

Shen KaiYue[1], Yan Da[1], Yin HuaQiang*[,1], He XueDong[1], Ma Tao[1], Fan hua[1]

1.Institute of Nuclear and New Energy Technology, Collaborative Innovation Center of Advanced Nuclear Energy Technology, Key Laboratory of Advanced Reactor Engineering and Safety of Ministry of Education, Tsinghua University, Beijing 100084, China

Email address:yinhuaqiang@tsinghua.edu.cn

Abstract. High-temperature gas-cooled reactor (HTGR) is the first kind of power plant with fourth-generation features in the world, and a large amount of graphite and carbon material is adopted as structural material and fuel matrix material, including boron-containing carbon (BC) and isostatic pressure graphite (IG-110). As a kind of porous material, the carbon material contains a certain amount of moisture, which will impose negative effects on reactor safety. Therefore, this paper studies the water vapor adsorption characteristics of carbon materials in HTGR. The adsorption isotherms of BC and IG-110 were obtained by experimental measurements. Experiments show that the moisture adsorption of IG-110 is much lower than that of BC at various relative humidity. The experimental data is analysed by the BET, GAB, D-A, Oswin adsorption models.

1. Introduction

High-temperature gas-cooled reactor (HTGR) is the first kind of power plant with fourth-generation features in the world, and a large amount of graphite and carbon material is adopted as structural material and fuel matrix material, including boron-containing carbon (BC) and isostatic pressure graphite (IG-110) [1]. As a kind of porous material, there are many micro pores in it, and therefore adsorbs moisture through pores from surroundings at room temperature. The specific surface area of carbon material is very high, and the main absorption type is physical absorption.

The moisture in carbon materials will be released into the helium coolant during the temperature-rise process of reactor. There is about 60t of graphite material used in the HTR-10[2], and almost 1000t in a single reactor core of HTR-PM[3]. Approximately 100t of graphite is used in AVR in Germany and 700kg of water is released during its power-rise and temperature-rise process. In HTR-10, about 130kg of water is released.

The chemical corrosion reaction between graphite and moisture exists in the high-temperature condition in an operating reactor, and therefore imposes bad influence on material strength and service life. Especially in the case of air and water leakage under nuclear power accident conditions, the corrosion is even more serious[4]. In order to decrease the corrosion, the reactor construction, fuel and graphite spheres and the space of primary system shall be heated and dehumidified. The limiting value of moisture content in helium coolant is $2cm^3/m^3$[5]. Therefore, study on the water vapor adsorption characteristics of carbon materials in HTGR is essential.

Numerous studies have investigated water vapor adsorption on various porous medium. With the different relative pressures, the adsorption forms of the adsorbates (water vapor) in the porous material structure are different. When the relative pressure is very low, water vapor molecules are first captured by the adsorption active center on the surface of the material, and then form monomolecular adsorption layer[6]. When the relative pressure rises to a certain value, the water vapor will undergo capillary condensation in the pore structure of the material [7]. As the further increases of relative pressure, the pore structure of the material will be gradually filled with water vapor.

The International Uniform of Pure and Applied Chemistry (IUPAC) classifies the pore sizes as follows: micropore (< 2 nm), mesopore (2-50 nm) and macropore (>50 nm). Different materials exhibit different adsorption isotherms corresponding to their pore sizes. Although different materials exhibit different adsorption isotherms, the adsorption isotherms of most materials can be classified into 5-6 types, as shown in Figure 1[8]. Type I is characteristic for purely microporous material, such as zeolite or activated carbon. Type II for pure nanoporous or macroporous material. Type III for the rare cases of nonporous materials with small interaction between adsorbent and adsorbate. Type IV for common mesoporous material, while type V presents relatively rare mesoporous materials, which is also due to the small interaction between adsorbent and adsorbate. Type VI is quite rare, the adsorption process is carried out step by step.

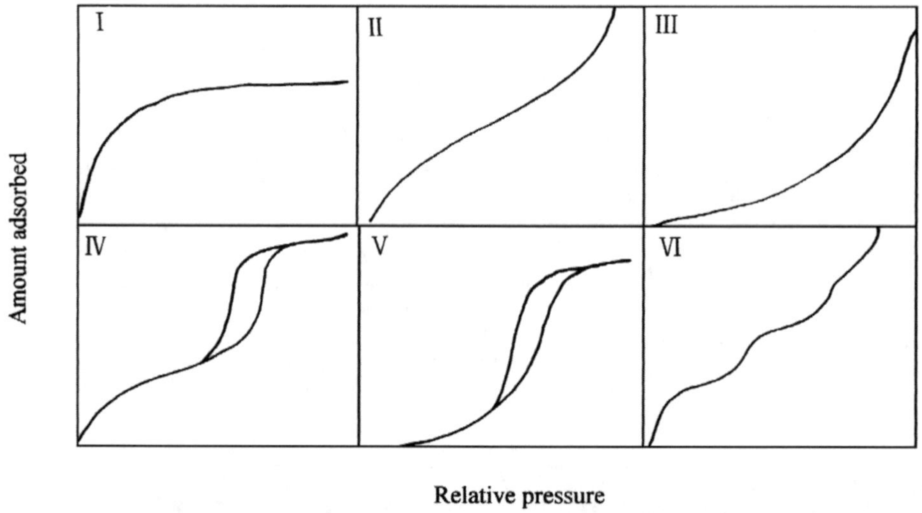

Figure 1. The original types of isotherms formulated by IUPAC.

In 1916, Langmuir proposed the theory of monolayer adsorption, which satisfactorily explained the type I adsorption isotherm [9]. He regarded the adsorption equilibrium as a dynamic equilibrium. That is, the gas or vapor molecule impacts the solid surface, and stays on the gas surface for a certain time, and then gets enough energy to break away from the surface and desorption occurs. Brunauer, Emmett and Teller put forward the theory of multilayer adsorption based on Langmuir equation in 1938, and deduced the BET equation [10], which has more general applicability. Later studies have found that for most materials, the assumptions of the BET model are not completely correct. What's more, this model is only applicable within the limited humidity range of 5 %RH to 35%RH. In 1946, Oswin developed an empirical model formula [11]. Although there are relatively few literature reports using the Oswin model equation, its fitting effect on the experimental data of food isothermal moisture absorption is better than the BET formula in most cases. In the 1980s, based on the Langmuir single-molecule adsorption equation and the BET equation, Guggenhern, Anderson, and De Boer introduced additional degrees of freedom and deduced the three-parameter GAB equation in which the adsorption layer is multimolecular layer [12]. The GAB equation, which apply to 0-90% RH, has been considered as the most general adsorption model formula since it was proposed.

For the carbon materials BC and IG-110 used in high temperature reactor, some studies have been carried out on their hygroscopicity. The results show that the equilibrium moisture absorption of BC is only about 0.1%.[13] The moisture absorption is closely related to the humidity of the environment, and the more moderate the equilibrium moisture content is, the greater the equilibrium moisture content is. These studies have some limitations, and haven't measured the complete adsorption isotherms of BC and IG-110 for water vapor. In this paper, the phenomena of water vapor adsorption and desorption of BC and IG-110 are detailed studied, and the water vapor adsorption isotherms of samples are measured.

2. Materials and Methods

2.1. Materials

The first high temperature gas-cooled experimental reactor HTR-10 is built in 2000 [14], which used more than 60 tons of carbon materials in its core. The BC material was provided by FANGDA Carbon New Material co., LTD, China, and IG-110 was provided by TOYO TANSO CO., LTD, Japan. The experimental samples are shown in Figure 2. The cylindrical samples are both 10mm in diameter and 5mm in height.

Pore parameters of BC and IG-110 measured by mercury intrusion method and Nitrogen adsorption method are shown in Table 1.

Table 1. Pore parameters of BC and IG -110.

Parameter	BC	IG-110	Unit
Size	D10*H5	D10*H5	mm
Weight	726.296	736.414	mg
Volume Density	1.7468	1.8259	g/cm^3
Apparent Density	2.0295	2.1560	g/cm^3
Porosity	13.93	15.31	%
Specific surface area	12.581	15.198	m^2/g
Most probable aperture	7930	1716	nm

2.2. Methods

The dynamic water vapor absorption instrument (DVS Adventure, Surface Measurement Systems, England) was used to the water vapor absorption and desorption test. The samples were first cleaned by ethanol, then dried in a drying oven at 250 ℃for 12 hours, and then cooled to environment temperature in a glass vessel containing silica gel desiccant. After drying, the samples were put into the climatic chamber of the DVS instrument and placed on the metal pallet, and the automatic balance will weigh pallet precisely at regular intervals. The parameters of the water vapor absorption test are shown in Table 2.

Table 2. Water vapor absorption test parameters.

Parameter	Value	Unit
Temperature	5-85	℃
Max. weighing capacity	1500	mg
Weighing precision	0.0001	mg
Relative humidity cycle	0-98 (5-60℃)	%
	0-85 (5-60℃)	%

For each experiment the sample of BC and IG-110 were used for water sorption analysis. Using the multistep technique adsorption isotherms for each sample were obtained at 20℃. Under each humidity, it is considered to have reached equilibrium when the mass change rate is less than 0.0002% within 15 minutes.

3. Experimental results and analysis

3.1. Modelling of sorption isotherms

The adsorption data is analysed for various adsorption models which include: （i）Brunauer–Emmett–Teller (BET)；（ii）Dubinin–Astakhov (D–A)[15]; （iii）Oswin （iv）Guggenhern, Anderson, and De Boer (GAB). The corresponding mathematical expression for adsorption models are given in Table 3.

Table 3. Adsorption isotherm models used for fitting of the experimental data.

Model	Mathematical expression	Estimated parameters	
BET	$M=\frac{M_m C(P/P_0)}{[1-(P/P_0)][(1-(P/P_0))+C(P/P_0)]}$	M_m, C	(1)
D-A	$M=M_0\exp\left[-\left(\frac{RT\ln(P/P_0)}{E}\right)^n\right]$	M_0, n, E	(2)
Oswin	$M=A\left[\frac{(P/P_0)}{1-(P/P_0)}\right]^B$	A, B	(3)
GAB	$M=\frac{M_m C_{GAB} K(P/P_0)}{[1-K(P/P_0)][(1-K(P/P_0))+C_{GAB}K(P/P_0)]}$	M_m, C_{GAB}, K	(4)

3.2. Fitting of adsorption isotherm

The values of obtained fitting parameters of samples of BC and IG-110 are given in Table 4. The parameters of each model are identified in the nomenclature. Linear technique is used to determine the model parameters and goodness of the fit by minimizing the mean relative percentage deviation modulus, E (%) which is defined as[16]:

$$E(\%)=100*\frac{1}{N}\sum_{i=1}^{N}\left[\frac{|M_e-M_c|}{M_e}\right] \tag{5}$$

Where N is the number of experimental points, M_e is experimental and Mc is calculated adsorption uptakes (g/kg).

Table 4. Adsorption isotherm models used for fitting of the experimental data.

Model	Estimated parameters	BC	IG-110
BET	M_m (g/kg)	0.0836	0.0161
	C	5.3449	3.7028
	R^2	0.9992	0.9996
	E (%)	13.47	20.25
D-A	M_0 (g/kg)	4.8563	1.2694

	n	0.3	0.3
	E(kJ/kg)	1.4319	1.2306
	R^2	0.9998	0.9999
	E (%)	4.43	7.24
Oswin	A	0.1503	0.0355
	B	0.6809	0.6670
	R^2	0.9998	0.9999
	E (%)	5.47	8.17
GAB	M_m (g/kg)	0.0985	0.0229
	C_{GAB}	4.1938	4.1973
	K	0.9555	0.9549
	R^2	0.9998	0.9999
	E (%)	3.92	5.45

R^2 is the coefficient of determination.

Figure 2 and Figure 3 show the experimentally measured adsorption isotherms data for the BC and IG-110 material respectively, together with their fits using the BET, GAB and D-A theoretical model. As seen in Figure 2, BET, GAB and D-A equation can fit the experimental data of BC well at low relative humidity, but in high relative humidity, the prediction accuracy of BET is obviously lower than that of GAB and D-A model.

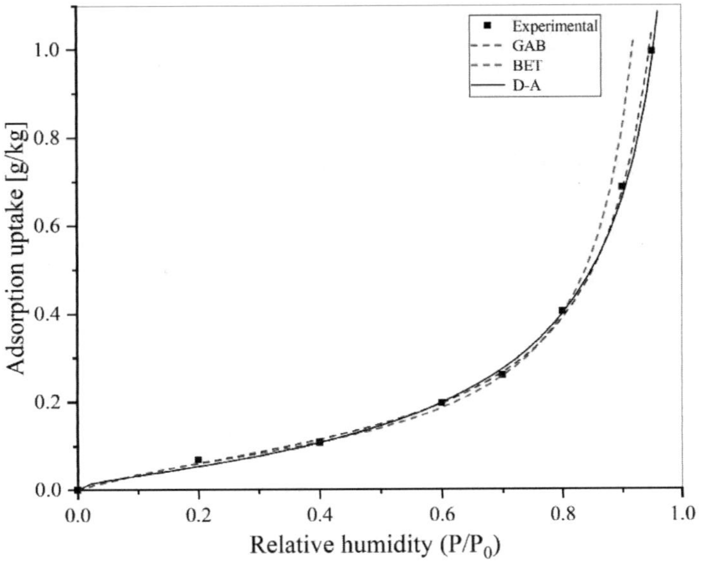

Figure 2. Fitting of experimental data for BC.

Figure 3 shows that the equilibrium adsorption capacity of IG-110 for water vapor is much lower than that of BC. The GAB and D-A equation can well simulate the trend of experimental data for IG-110, which is much better than the BET equation.

4. Conclusions

The adsorption isotherms of BC and IG-110 were obtained by experimental measurements Adsorption isotherms of water vapor onto two porous materials BC and IG-110 have been experimentally measured using the dynamic water vapor absorption instrument (DVS Adventure) at adsorption temperature 20°C. Experiments show that the moisture adsorption of IG-110 is much lower than that of BC at various relative humidity. The adsorption data is analyzed by various adsorption models which include BET, GAB, D-A, Oswin model. The GAB model gives the good fit for the relative humidity range of 0–0.95. The mean relative percentage deviation modulus is found 3.92 and 5.45 % for BC and IG-110, respectively.

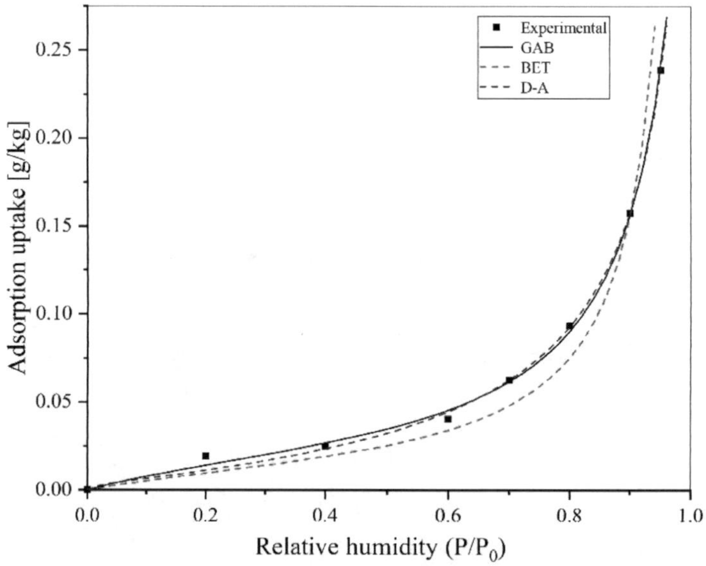

Figure 3. Fitting of experimental data for IG-110.

References

[1] Zhang, Zuoyi, and Suyuan Yu. "Future HTGR developments in China after the criticality of the HTR-10." Nuclear Engineering and Design 218.1-3 (2002): 249-257.e

[2] XU Wei, SHI Lei, ZHENG Yanhua, LIU Peng. Effect of Porosity on Oxidation of Nuclear Grade Graphite [J]. Atomic energy science and technology, 2016, 50(3): 471-476.

[3] Dou Jinyuan, Yin Huaqiang. Preliminary investigation into the dehumidification model of HTR-PM initial core [J]. The 14th National Reactor Thermal Fluid Academic Conference. 2015.

[4] Yu Xinli, Yang Xiaoyong, Yu Suyuan. Rsearch Status of Oxidation of Graphitein in HTR in Case of Leakage of Water and Air [J]. Nuclear Power Engineering, 2006: 313-315.

[5] YU Xin-li, LUO Xiao-wei, YU Su-yua. Simulation of Oxidation in HTR-10 Core [J]. Nuclear Power Engineering, 2010, 31(2): 81-97

[6] Fytianos K, Voudrias E, Kokkalis E. Sorption—desorption behaviour of 2,4-dichlorophenol by marine sediments[J]. Chemosphere, 2000, 40(1): 3-6.

[7] Yakout S M, Daifullah A A M, El—Reefy S A. Adsorption of naphthalene, phenanthrene and pyrene from aqueous solution using low—cost activated Carbon derived from agricultural wastes[J]. Adsorption Science and Technology,201 3, 3 1(4): 293—302

[8] Thommes, Matthias . "Physisorption of gases, with special reference to the evaluation of surface area and pore size distribution (IUPAC Technical Report)." Pure & Applied Chemistry

38.1(2016):25-25.
[9] Langmuir, Irving. "THE CONSTITUTION AND FUNDAMENTAL PROPERTIES OF SOLIDS AND LIQUIDS." Journal of the Franklin Institute 184.5(1916):102–105
[10] Brunauer, Stephen , P. H. Emmett , and E. Teller . "Adsorption of Gases in Multimolecular Layers." Journal of the American Chemical Society 60.2(1938):309-319.
[11] Oswin, C.R.: The kinetics of package life. III. The isotherm. J. Soc Chem. Ind. 65, 419–421 (1946)
[12] Mihoubi, D., Bellagi, A.: Thermodynamic analysis of sorption isotherms of bentonite. J. Chem. Thermodyn. 38, 1105–1110(2006)
[13] Jun, L. I., et al. "Moisture Absorption Property and Diffusion Coefficient of Carbon Material with Boron Carbide in HTGR." Atomic Energy Science & Technology (2018).
[14] Lohnert, G. H . "Technical design features and essential safety-related properties of the HTR-module." Nuclear Engineering & Design 121.2(1990):259-275.
[15] Dubinin, M.M., Astakhov, V.A.: Development of the concepts of volume filling of micropores in the adsorption of gases and vapors by microporous adsorbents. Bull. Acad. Sci. USSR Div.Chem. Sci. 20, 3–7 (1971)
[16] Sultan M , El-Sharkawy I I , Miyazaki T , et al. Insights of water vapor sorption onto polymer based sorbents[J]. Adsorption-journal of the International Adsorption Society, 2015, 21(3):205-215.

Acknowledgments
This work was supported by the National Natural Science Foundation of China (NSFC)(Grant No. 11875176), and the Chinese National S&T Major Project (Grant No. ZX06901).

IOP Conf. Series: Materials Science and Engineering 1150 (2021) 012009 doi:10.1088/1757-899X/1150/1/012009

The deterioration of 100 years old coated steel bridges

C. Paglia, S. Antonietti and C. Mosca

University of Applied Sciences of Southern Switzerland, DACD, 6850 Mendrisio, Switzerland.

E-mail: christian.paglia@supsi.ch

Abstract. Steel bridges have been widely built from the beginning of the XIX century. The structures were protected from the uniform atmospheric corrosion through Lead-containing coatings. These exhibited over the decades a very good resistance to degradation. Generally, a similar reduced corrosion was also observed for the rivets, which linked the structural elements. The degradation was mainly located in the regions directly exposed to rainfall or water flux. The formation of craters or exfoliation was only locally found. The enrichment of organic deposits promoted the permanence of humidity, which in turn, had an adverse effect on the steel conservation. Generally, the total coating thickness slightly decreased, and the long-therm protection was no longer attained. The content of Lead has a detrimental effect on the environment. Thus, the restoration works must consider the pollutant effect during its elimination and disposal.

1. Introduction

Corrosion is an electrochemical process that occurs between the metal surface and the environment. Water or humidity, gases such as Oxygen, acids or bases and salts, as for instance NaCl, promote the phenomena [1]. Corrosion protection coatings are developed and classified according to the environmental atmosphere corrosivity [2]. The classes are depending on the aggressivity, starting from a class C1 with a very low risk, to a conventional C3 medium urban environment, ending with a very high risk for industrial areas or coasts, as well as off-shore zones with an aggressive atmosphere. The system type and the layers are variable. The durability of the systems is influenced by the condition and support preparation, the coating system, the quality and the application condition, as well as the exposition [3]. The modern coating systems are divided in classes starting from a low durable system, that may reach 2 to 5 years, followed by a medium, with 5 to 15 years durability and high classified coatings, that may last beyond 15 years. Nonetheless, the durability appears to be indicative. In the past, an anti-rust orange base layer followed by an upper grey Lead-containing one were applied on steel elements. In particular, bridges and structural bars. Although in the early 1900s health hazard risk were identified, Lead was used as a paint pigment, giving a thick, resistant coatings. They do not crack during temperature variation and are resistant to wear. Furthermore, the chemical behavior also provides a good corrosion resistance [4, 5]. The goal of the present work is to characterized the main steels used and to outline the main degradation defects on old steel bridges coated with Lead-containing materials.

2. Experimental

The steel was characterized with tensile tests at room temperature [6]. The impact resistance was measured with Charpy tests [7, 8]. The specimens were oriented perpendicular to the longitudinal axes with a V notch of 45°. The tests were carried out at 20 ± 2 °C and -20 ± 2 °C with slightly reduced sections. Therefore, the measured values need to be taken as indicative. The specimens did not comply with the conventional determination of the äquivalent carbon. Therefore, the weldability was alternatively tested according to a European norm [9]. The Vickers hardness across the welded profiles did not exhibit an increase in the hardness along the weldments. That means a good steel weldability. The coating thickness was measured with a magnetic induction method by using a dualscope. A very large investigation of the bridge elements was done by visual inspection. In this concern, appropriate safety concern must be taken into account (Fig. 1).

Figure 1. Visual inspection on a steel bridge.

3. Results and discussion

The bridges investigated in this work were mainly built in the first years of the XIX century. The steels are usually characterized with the respect to the mechanical properties, especially by tensile tests. Although the steel type may vary depending on the structure, the mean necking limit was around 292 ± 5.4 N/mm^2. This is typical for a S235 steel type. The mean tensile strength was 387 ± 1.9 N/mm^2 and the mean strain was 37 ± 1.0 %. The impact behavior was measured by means of Charpy tests. The results exhibited a clear difference in the adsorbing energy at the different temperatures (Tab. 1).

Table 1. Charpy test results of the specimens at different temperatures.

Denominazione campione	Temp. [°C]	S_0 [mm^2]	Energia [J]	KV [J/mm^2]
L1	20	50	66	1.32
L2	20	50	66	1.32
L3	20	50	40	0.80
C4	20	50	45	0.90
C5	20	50	38	0.76
L6	-20	50	2	0.04
C7	-20	50	2	0.04
C8	-20	50	2	0.04
C9	-20	50	2	0.04
L10	-20	50	2	0.04

The adsorbed energy varied from 38 to 66 J at 20 ± 2 °C and was stabilized at 2 J for the specimens tested at -20 ± 2 °C. The fracture mode was ductile or brittle depending on the temperature. The chemical analysis exhibited a steel with low Carbon content (0.052 %). The steel for the rivets had a higher Carbon content (0.130 %) and also the Copper (Cu 0.015 %) was high. A relatively high Sulphur content was also detected (S 0.094 %). The low Carbon content can be correlated with a good ductility. The Sulphur lowers the ductility. The Phosphorus (P 0.044 %) implements the brittleness at low temperature.

In general, the mean coating thickness of the steel elements exhibited values ranging from 213.7 +/- 49.5 μm to 99 +/- 13.2 μm in the relatively good conserved zones. The typical environment and the corrosion products analysis of acqueous solutions indicated a low chloride and sulphate concentration, thus reducing the level of contamination. The steel coatings exhibited an anti-rust layer (orange) and a superficial cover grey colored (Fig. 2). A semi-quantitative X-ray fluorescence analysis exhibited a Lead (Pb) content up to 38% by mass.

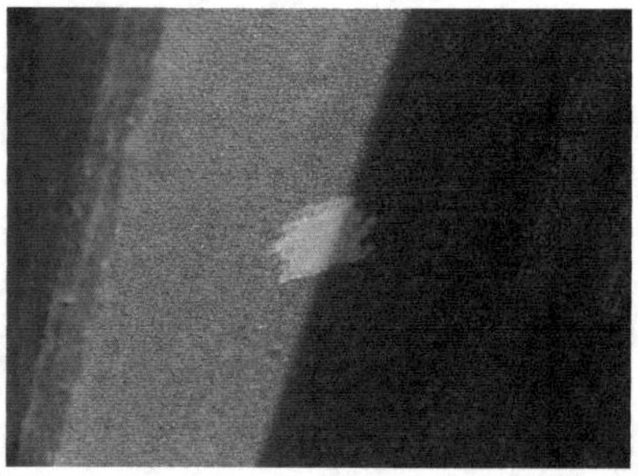

Figure 2. Steel coating with an anti-rust (orange) and an upper layer (grey).

Generally, after almost a century of atmospheric exposure, the coating exhibited a relatively good conservation state. In general, uniform corrosion is present. By mechanical removing of the corrosion layer, there is an absence of a significant degradation (Fig. 3 left). Sporadic increased degradation remains at a superficial level (Fig. 3 right binocular lenses image).

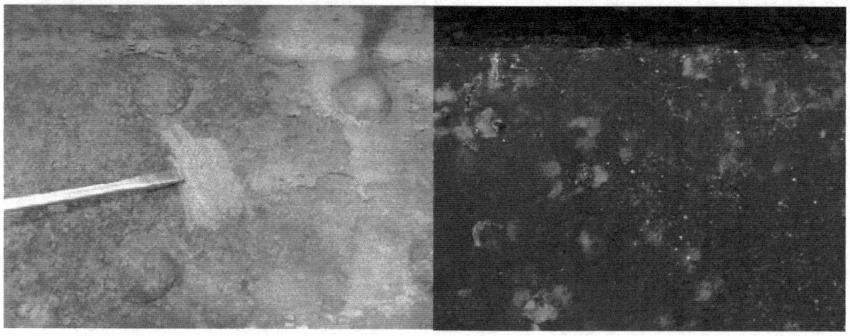

Figure 3. Uniform corrosion of the coated steel elements.

Part of the elements protected from direct exposition to atmospheric agents exhibit a decreased corrosion (Fig. 4 left). Coatings are relatively still present, in particular on the elements sheltered from rain. Metal degradation is present along water pathways as well (Fig. 4 right).

Figure 4. Correlation between corrosion and direct-indirect exposition to the atmospheric agents.

The rivet heads generally exhibit a slight uniform corrosion (Fig. 5 left). A similar reduced degradation is also present along the rivets, once extracted from the holes (Fig. 5 center). On the other hand, these parts of the structure can be especially critical with respect to the environmental degradation. In the bridge support regions, the corrosion may exhibit an advanced corrosion stage, with thickness loss of the metal elements, such as for instance the rivet heads (Fig. 5 right).

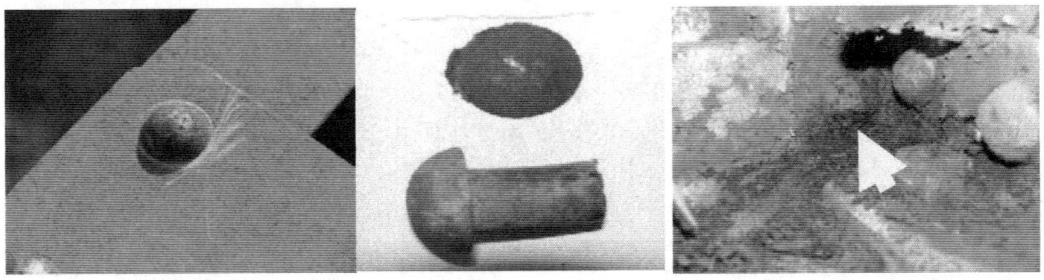

Figure 5. Corrosion stage of the rivets and rivet heads.

Additional critical parts to be carefully considered are the concrete embedded steel regions, which may be affected from advanced corrosion (Fig. 6 left). While in some cases the rivets comes out from the bridge elements, endangering the stability of the elements (Fig. 6 right).

Figure 6. Steel bridge support region in contact with the cementitious material (left). Rivets detachment (right).

Enriched organic and inert deposits may act as humidity catcher, thus promoting corrosion. These deposits are present at the base part of the structures, often when the steel elements are anchored to the basement (Fig. 7 left). Some elements next to each other may create empty narrow spaces, that promote enrichments of deposits (Fig. 7 right) and humidity. Therefore, it is necessary to periodically clean these zones, to avoid corrosion.

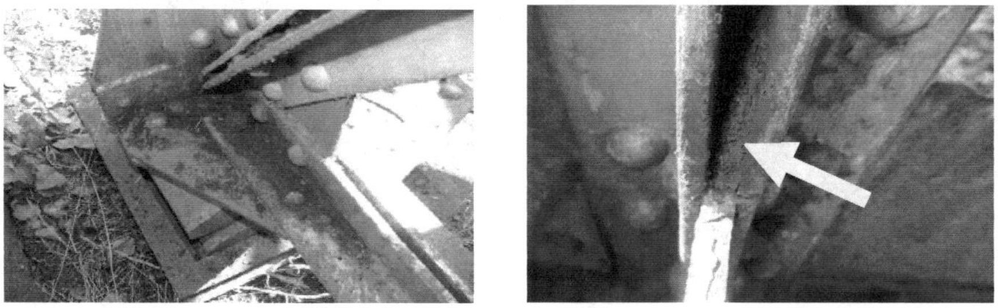

Figure 7. Organic deposits along the steel elements.

The deposit points have to be cleaned to avoid permanent presence of humidity and corrosion. In points subjected to major fatigue, the corrosion products need to be eliminated in order to investigate the presence of microcracks.

Conclusions

The steel bridges built over a century ago often exhibit mechanical parameters of steel type S235. The conservation state of the Lead-enriched coatings is generally good. Uniform atmospheric corrosion takes place on a superficial level. Supports may exhibit an advanced corrosion stage with the formation of craters and steel thickness reduction. Exfoliation is caused by the enrichment of organic deposits as well as the long-therm permanence of humidity. The steel supports embedded in concrete might be critical with respect to corrosion. In this latter case, the deterioration must be investigated with additional techniques, such as corrosion potential, pulse measurements and radar. The Charpy tests indicated the structure to be subjected to brittle fracture at low temperature in case of impact load. Generally, the steels exhibit a low Carbon content, which correlates with a good ductility and weldability. Nonetheless, the coatings thickness is no longer sufficient to guarantee future protection. Therefore, the old coatings need to be eliminated and accurately treated, considering the Lead content and the environmental issue to be respected.

Acknowledgements

The authors would like to thank Peter Heimgartner from Qualitech AG for the discussions.

References

[1] E. McCafferty, Introduction to corrosion science, Springer, 2009.
[2] Norm ISO 12944-2:2017, Paints and varnishes, Corrosion protection of steel structures by protective paint systems, Part 2: Classification of environments, 2017.
[3] Norm ISO 12944-1:2017, Paints and varnishes, Corrosion protection of steel structures by protective paint systems, Part 1: General introduction, 2017.
[4] Federal highway administration research and technology, U. S. Department of Transportation, Lead abatment on bridges and steel structures, 2016.

[5] Lead-containing paints and coatings, Preventing exposure in the construction industry, Worksafe BC, 2011.
[6] Norm ISO 6892-1, Metallic materials — Tensile testing — Part 1: Method of test at room temperature, 2019.
[7] Norm ISO 148-1, Metallic materials — Charpy pendulum impact test — Part 1: Test method, 2016.
[8] Norm ISO 148-2, Metallic materials — Charpy pendulum impact test — Part 2: Verification of testing machines, 2016.
[9] ISO 6507-1, Metallic materials — Vickers hardness test — Part 1: Test method, 2018.

IOP Conf. Series: Materials Science and Engineering 1150 (2021) 012010 doi:10.1088/1757-899X/1150/1/012010

Chemical and microstructural characterization of ash from the Ubinas and Sabancaya volcanoes in Peru

Danny Tupayachy-Quispe[1], Fredy Apaza[2], Rossibel Churata[3], Jonathan Almirón[3], Marián Hermoza-Gutiérrez[1], Francisco Velasco[4], Manuel Torres-Carrasco[4]

[1]Laboratorio de Ciencia de los Materiales, Universidad Católica de Santa María, Umacollo, Arequipa, Perú.
[2]Geological, Mining and Metallurgical Institute - INGEMMET, San Borja, Lima, Perú.
[3]Escuela Profesional de Ingeniería Ambiental, Facultad de Ingeniería de Procesos, Universidad Nacional de San Agustín de Arequipa, Cercado, Arequipa, Perú.
[4]Materials Science and Engineering Department, IAAB, Universidad Carlos III de Madrid, Leganés, Madrid, Spain.

E-mail: dtupayachy@ucsm.edu.pe

Abstract. In this study, volcanic ash from Ubinas and Sabancaya volcanoes in Peru were characterized by particle size, scanning electron microscopy coupled with energy dispersive X-ray spectrometry (SEM/EDX), ICP MASS spectrometer, X-ray diffraction and specific gravity. The combined application of those techniques for the analysis provided complementary information, i.e., their elemental compositions and morphology. The results showed that the volcanic ashes (VA) have a particle size lower than 600 μm with a dominant size over 100 μm. Their mineralogical characteristics involve silica polymorphs, vitreous nature of silica and silicate particles, showing chemical heterogeneity within the micrometer sized particles and diverse morphology. The chemical composition of the ash is mainly SiO_2, Al_2O_3, Na_2O and CaO, with a high content of silica ranging from 57.78 to 66.14% and alumina ranges from 13.81 to 17.56%. The concentration of potential pollutant elements determined in products from Ubinas and Sabancaya volcanoes, namely As, Ba, Co, Cr, Cs, Fe, K, Mn, Mo, Sb, Sc, Sr, Th, Ti, U, V, and Zn, are reported. The different mineralogical characteristics among the studied volcanic samples would be related to their association with different volcanic eruptions. The chemical compositions and microstructure indicate that ashes have an origin and formation of andesitic and dacitic type.

1. Introduction

The Central Volcanic Zone (CVZ) is part of the Andean Volcanic Belt and it extends from Peru to Chile for approximately 1500 km in length. This volcanic arc, related to a segment of the subduction zone of the Nazca plate below South America, is house of 69 Holocene volcanoes [1]. Eight active volcanoes are located in the southern Peru segment of the CVZ, being Ubinas and Sabancaya the most active volcanoes currently among them [2].

The Ubinas volcano is located 80 km southeast of the city of Arequipa, in the province of Sanchez Cerro. It has an approximate height of 5672 meters above sea level. Since its last Plinian eruption in A.D.1000-1160, and until 1996, the volcano has experienced to 24 erupting episodes [3]. In 2013, a new eruptive cycle began, characterized by an initial phreatic phase followed by vulcanian explosions emitting juvenile material of andesitic composition [4]. In 2014, the characterization of the juvenile magma of the Ubinas volcano emitted in the eruptive activity of 2006 was reported, resulting in a homogeneous andesitic composition of the rocky massif (56.7-57.6% by weight of SiO_2), which belongs to a series of medium-high K chalcoalkalines [5].

Sabancaya is located 80 km northwest of the city of Arequipa in the district of Achoma, province of Caylloma, at 5976 meters above sea level [6]. The first records of eruptions at Sabancaya volcano date back to 1750-1780 [7]. Activity was also reported in 1986, with volcanic ash emissions culminating in explosive phases in the 1990s with volcanic clouds exceeding 7 km in height [8]. Juvenile material was andesitic to dacitic [8]. In 2012, degassing activity increased and, in 2016, a new eruptive phase began, with abundant ash emissions around the volcano [9].

The explosive eruptions in Ubinas and Sabancaya spew abundant volcanic ash, causing damage to various environmental sources [10]. However, previous studies have used ash from the Ubinas volcano for the development of geopolymers, which have a potential use in construction [11]. Also, there are other studies where volcanic ash has been used as flame retardants with polypropylene matrices [12].

The volcanological evolution and petrology of magma for both volcanoes has been previously described [5], but there is not much information about the physical characteristics and chemical composition of ash from Ubinas and Sabancaya volcanoes.

The aim of this work is to provide new information on the chemical characteristics, mineralogy and morphology of the volcanic ash of Sabancaya and Ubinas, through particle distribution analysis, determination of specific gravity, scanning electron microscopy, ICP MASS spectrometry and X-ray diffraction tests.

2. Methodology

2.1. Sampling

Three representative samples of volcanic ash were taken from the bases, two from the flanks and one from the craters of the Ubinas and Sabancaya volcanoes, always considering areas with large deposits of erupted material. Table 1 shows the coding and geographic location of the samples from each volcano.

Table 1. Sampling points in the Ubinas and Sabancaya volcanoes.

Sampling area	Ubinas volcano				Sabancaya volcano			
	Sample Code	Coordinates			Sample Code	Coordinates		
		Latitude	Longitude	Height		Latitude	Longitude	Height
Base	UB1	297608	8196158	4600	SB1	203247	8246403	4440
	UB2	295537	8187927	4384	SB2	201068	8246586	4488
	UB3	300699	8191317	4354	SB3	198916	8256988	4883
Flank	UF1	294939	8191824	4850	SF1	195364	8250857	5277
	UF2	299284	8190784	4677	SF2	195150	8248396	5074
Crater	UC	297429	8192472	5396	SC	194104	8252701	5828

2.2. Characterization tests

The samples were dried at 102±2 °C for 5 h for later characterization. Ash particle size distribution was determined using a Malvern MASTERSIZER 2000 laser particle scattering particle analyzer. Specific gravity was measured with a Micromeritics AccuPyc 1330 helium pycnometer.

ZEISS EVO 15 scanning electron microscope (SEM) was used to perform morphology analysis of the ashes using a backscattered electron (BSE) detector. Analysis was carried out using 15 kV voltage. The selected samples were initially sieved using meshes of 149 μm and 74 μm. A sample fraction was then placed in a sample holder using a carbon tape, and finally coated with carbon.

Inductively coupled plasma (ICP) atomic emission spectroscopy were performed on the ICP MASS SPECTROMETER equipment, brand Perkin Elmer Nexion 300D. An acid attack with HNO_3 and HCl was applied to dissolve the sample for 12 h. The samples were taken to high plasma temperatures between 6000 °C and 7000 °C to be analyzed, in order to eliminate traces of water and organic vapors. X-ray diffractograms were obtained using a Bruker diffractometer (D8 Avance Davinci) with CuKα radiation (λ = 0.1542 nm) operated at 40 kV and 40 mA. Bragg angles ranging 10 ° to 80 ° were investigated at a scanning speed of 1 ° min^{-1}.

3. Results and discussion

3.1. Particle size distribution

Results are illustrated both as volume percentages for individual grain size fractions and cumulative form (Figure 1). SC, SF1 and UC volcanic ash samples are the most enriched in finer ash components. SC and SF1 samples, having a size distribution between 5 and 600 μm, are fresh-fallen juvenile ashes of the 2016-2018 Sabancaya eruptive sequence (collected as they deposited).

UC sample, deposited during the recent Ubinas eruptions in 2013-2016, shows a bimodal distribution, with peaks at approximately 40 and 600 μm in diameter. In contrast, UF1, UB1 and SB1 samples have a homogeneous size distribution. These samples are affected by the weather after deposition, with a range between approximately 100 and 600 μm.

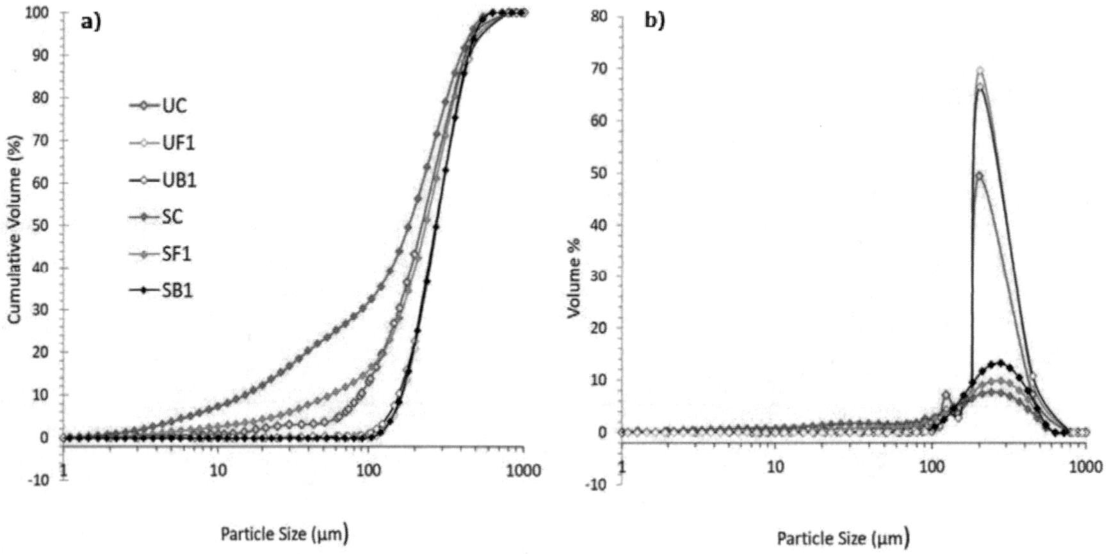

Figure 1. Average results (3 runs) of grain size distributions expressed as volume percentages of grain size (a) and cumulative volume (b) of Ubinas and Sabancaya ashes.

3.2 Specific gravity

Based on the specific gravity results (Table 2), the volcanic ash samples from Ubinas and Sabancaya can be classified in the light material category, considering that the primary mineral composition of volcanic ash is typically characterized by dividing minerals into light (specific gravity < 2.8-3.0 g/cm^3) and heavy (specific gravity > 2.8-3.0 g/cm^3) mineral categories [13]. Light minerals dominate in volcanic ash with abundances typically ranging between 70 and 95% [13]. Within the light mineral category, the relative mineral abundance typically decreases in the order: non-colored volcanic glass > plagioclase feldspars > silica minerals (quartz, cristobalite and tridymite) ≈ mica.

Table 2. Specific gravity of volcanic ash Ubinas and Sabancaya volcanoes.

	Volcanic ash from Ubinas						Volcanic ash from Sabancaya					
Sample code	UB1	UB2	UB3	UF1	UF2	UC	SB1	SB2	SB3	SF1	SF2	SC
Specific gravity (g/cm^3)	2.72	2.77	2.74	2.85	2.67	2.70	2.74	2.71	2.72	2.70	2.71	2.68

3.3. Scanning electron microscopy

Upon SEM analysis, the samples appear composed predominantly of angular lithics, plagioclase, pyroxene, biotite, and irregular glass fragments [14]. It was observed that the samples have no defined shape. Shapes vary from irregular to sub-rounded, to stretched forms with angular surfaces; others clasts show distortion of vesicles and highly irregular shapes (Figure 2).

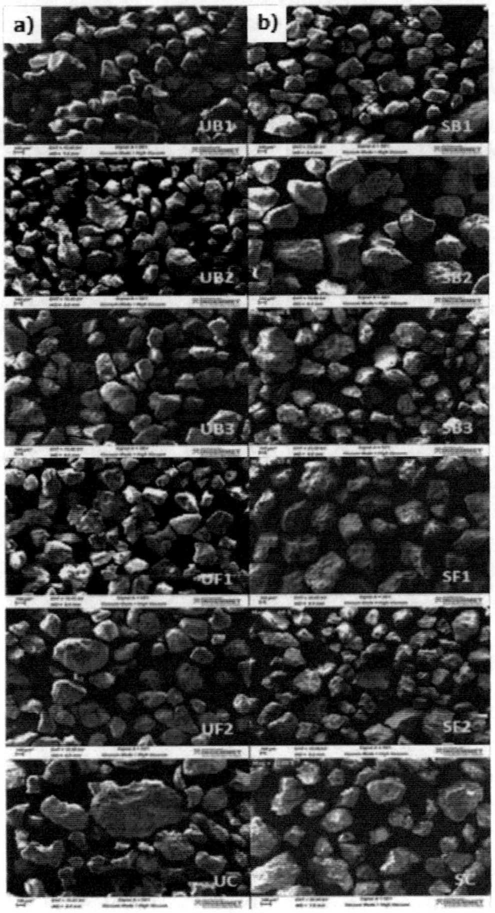

Figure 2. Ubinas (a) and Sabancaya (b) volcanoes ash morphology.

Ubinas ash samples vary considerably with sampling location. The UC sample, corresponding to the closest sampling site to the crater, exhibits the greatest number of particles with angular edges (Figure 2). There are not big differences between SC and UC morphologies. Moreover, the ash recovered from the sedimentary SB2 (for instance) has a particle size and morphology similar to SC, and thus corresponding to fresh ash.

In particular, the shape of the volcanic particles can provide important information about the conditions in the volcanic duct during particle formation. The characteristics of the sampling points have similar characteristics in both volcanoes, with a process of magma fragmentation [19,20].

The microstructure of the ash has angular and slightly rounded characteristics. It is composed mainly of glass with non-consolidated matrix, plagioclase crystals and pyroxene and little lithics. The observations in SC sample show the presence of hydrothermalized material. The average particle size is 22 and 280 um, however they present a variable size. In the SEM analysis with dispersive energy, it was found that the particles smaller than 15 μm represent 10% of the total sample.

3.4 Inductively coupled plasma atomic emission spectroscopy

Samples collected at Ubinas and Sabancaya have not only similar grain-size and morphological characteristics (Figure 1 and Figure 2), but also similar chemical composition. The composition of the volcanic particles (mainly of the fine ash) provides information on fragmentation, transport and deposition processes, as well as the physical properties of erupting magma. Table 3, where the results for major components are listed, indicates silica contents between 58% and 63% for Ubinas and 57% and 66% for Sabancaya.

Table 3. Chemical composition of volcanic ash from Ubinas and Sabancaya volcanoes.

Code	Al_2O_3 %	CaO %	Fe_2O_3 %	K_2O %	MgO %	MnO %	Na_2O %	P_2O_5 %	SiO_2 %	TiO_2 %	LOI*
UB1	14.80	5.80	7.69	2.63	4.17	0.10	3.52	0.37	58.70	1.14	0.87
UB2	14.58	5.17	6.23	2.15	3.15	0.08	2.94	0.30	61.69	1.11	2.34
UBF3	13.81	4.29	6.94	1.87	3.33	0.09	2.72	0.38	61.16	1.11	3.41
UBF1	14.34	4.82	6.18	2.23	2.92	0.08	2.91	0.30	61.80	1.19	2.39
UBF2	14.93	4.94	6.77	2.77	2.03	0.06	3.81	0.37	59.34	1.01	3.38
UC	14.77	4.24	5.90	2.49	2.32	0.07	2.87	0.34	62.92	1.15	2.11
SB1	15.84	6.21	6.69	2.02	4.27	0.09	3.52	0.24	57.78	0.99	1.26
SB2	14.64	3.80	5.33	2.48	2.17	0.06	3.67	0.30	66.14	0.91	0.41
SB3	15.66	4.62	4.91	2.14	2.98	0.06	3.36	0.32	63.30	0.84	1.21
SBF1	14.71	4.98	5.79	2.63	3.03	0.07	3.73	0.27	62.58	1.02	0.66
SBF2	16.26	3.85	5.37	2.47	2.23	0.06	3.68	0.32	63.97	0.93	0.60
SC	17.86	4.79	5.35	2.29	2.16	0.06	3.69	0.32	61.87	0.91	0.37
M.D.L.**	0.6	0.5	0.05	0.1	0.1	0.1	0.3	0.1	0.2	0.1	0.2

*Lost of ignition

**Minimum detection level

The concentration of elements determined in products from the Ubinas and Sabancaya volcanoes, namely Al, As, Ba, Ca, Co, Cr, Cs, Fe, K, Mn, Mo, Na, Sb, Sc, Sr, Th, Ti, U, V and Zn are reported in Table 4. The abundances of rare earth elements (REE) La, Ce, Nd, Sm, Eu, Tb, Dy, Yb, Lu and other relevant trace elements Hf, Rb, Ta, W and Zr are also reported.

IOP Conf. Series: Materials Science and Engineering 1150 (2021) 012010 doi:10.1088/1757-899X/1150/1/012010

Table 4. Elemental composition of the Sabancaya and Ubinas volcanic ashes.

			UBINAS						SABANCAYA					
		M.D.L.	UB1	UB2	UB3	UF1	UF2	UC	SB1	SB2	SB3	SF1	SF2	SC
Ag	ppm	0.5	<0.5	<0.5	<0.5	<0.5	<0.5	<0.5	<0.5	<0.5	<0.5	<0.5	<0.5	<0.5
Al	%	0.01	7.83	7.72	7.31	7.59	7.90	7.82	8.38	7.75	8.29	7.79	8.61	9.45
As	ppm	5	10	<5	6	<5	8	6	<5	<5	<5	<5	7	9
Ba	ppm	5	849	941	870	906	924	958	816	879	844	947	892	928
Bi	ppm	10	<10	<10	<10	<10	<10	<10	<10	<10	<10	<10	<10	<10
Ca	%	0.01	4.15	3.69	3.07	3.45	3.53	3.03	4.44	2.72	3.30	3.56	2.75	3.42
Cd	ppm	1	<1	<1	<1	<1	<1	<1	<1	<1	<1	<1	<1	<1
Co	ppm	1	27	19	21	20	15	15	24	15	16	20	15	17
Cr	ppm	1	61	67	59	52	26	45	97	31	35	79	38	38
Cu	ppm	2	27	35	36	32	23	30	48	30	35	37	32	44
Fe	%	0.01	5.38	4.35	4.85	4.32	4.73	4.12	4.68	3.73	3.43	4.05	3.76	3.74
K	%	0.01	2.18	1.79	1.55	1.85	2.30	2.07	1.68	2.06	1.77	2.18	2.05	1.90
La	ppm	2	43	54	41	52	41	50	56	40	38	57	41	41
Li	ppm	2	9	6	8	4	6	4	8	13	12	13	13	13
Mg	%	0.01	2.51	1.90	2.01	1.76	1.22	1.40	2.57	1.31	1.80	1.82	1.34	1.30
Mn	%	0.01	0.08	0.07	0.07	0.06	0.05	0.06	0.07	0.05	0.04	0.06	0.05	0.05
Mo	ppm	2	7	2	5	2	5	3	<2	<2	4	2	8	<2
Na	%	0.01	2.61	2.18	2.02	2.16	2.83	2.13	2.61	2.72	2.49	2.77	2.73	2.74
Ni	ppm	5	35	22	28	18	12	13	46	22	21	39	24	25
P	%	0.01	0.16	0.13	0.17	0.13	0.16	0.15	0.10	0.13	0.14	0.12	0.14	0.14
Pb	ppm	5	5	10	7	10	9	<12	<5	5	7	7	6	6
Sb	ppm	2	<2	<2	<2	<2	<2	<2	<2	<2	<2	<2	<2	<2
Si	%	0.01	27.43	28.84	28.59	28.89	27.74	29.41	27.01	30.92	29.59	29.26	29.90	28.92
Rb	ppm	2	49	54	51	59	57	68	51	76	63	78	70	66
Sr	ppm	2	788	850	756	789	759	748	872	761	748	817	765	824
Ti	%	0.01	0.68	0.67	0.67	0.71	0.61	0.69	0.59	0.55	0.50	0.61	0.56	0.54
Tl	ppm	5	<5	<5	<5	<5	<5	<5	<5	<5	<5	<5	<5	<5
V	ppm	10	205	152	164	152	145	138	148	131	132	152	134	141
W	ppm	5	<5	<5	<5	<5	<5	<5	<5	<5	<5	<5	<5	<5
Y	ppm	5	14	12	14	13	12	13	12	11	10	11	10	10
Zn	ppm	1	114	86	99	84	81	82	93	88	94	98	88	91
Zr	ppm	5	127	198	158	219	157	236	182	159	158	191	145	146
Be	ppm	0.6	1.6	2.3	1.6	2.2	1.7	2.2	2.5	2.2	2.1	2.3	2.0	2.1
Ce	ppm	0.5	70.0	73.5	71.9	72.8	74.7	72.3	71.9	71.8	68.6	72.1	70.8	72.6
Cs	ppm	0.05	1.59	1.89	1.91	1.92	2.03	2.21	1.22	1.88	1.86	1.85	1.77	1.74
Dy	ppm	0.1	3.4	3.0	3.1	3.2	3.1	3.0	3.0	2.6	2.4	2.7	2.6	2.7
Er	ppm	0.1	1.6	1.1	1.4	1.2	1.5	1.1	1.2	1.1	1.1	1.0	1.1	1.1
Eu	ppm	0.1	1.7	1.3	1.6	1.2	1.7	1.2	1.2	1.5	1.4	1.1	1.4	1.6
Ga	ppm	0.3	21.2	21.1	20.5	20.3	21.6	19.8	22.1	21.8	21.5	20.7	22.0	23.5
Gd	ppm	0.1	4.7	4.3	4.4	4.3	4.4	3.9	4.4	4.0	3.9	4.0	3.9	4.1
Hf	ppm	0.2	4.7	5.6	4.7	6.0	5.6	6.1	5.3	5.0	5.3	5.4	4.3	5.0
Ho	ppm	0.1	0.6	0.4	0.6	0.5	0.6	0.4	0.5	0.4	0.4	0.4	0.4	0.4
In	ppm	0.2	<0.2	<0.2	<0.2	<0.2	<0.2	<0.2	<0.2	<0.2	<0.2	<0.2	<0.2	<0.2
Lu	ppm	0.15	0.18	<0.15	<0.15	<0.15	0.17	<0.15	<0.15	<0.15	<0.15	<0.15	<0.15	<0.15
Nb	ppm	0.5	9.3	9.8	9.5	10.3	11.8	11.0	6.6	7.0	6.8	6.7	7.0	7.3
Nd	ppm	0.1	34.4	26.1	33.7	25.4	35.2	24.9	26.2	32.9	31.2	25.4	33.4	34.5
Pr	ppm	0.1	8.3	7.7	8.5	7.7	8.7	7.5	7.7	8.3	7.9	7.7	8.3	8.6
Sm	ppm	0.1	6.2	5.0	5.7	4.9	6.0	4.7	5.2	5.6	5.1	5.0	5.7	5.7
Ta	Ppm	0.5	<0.5	0.8	<0.5	0.8	0.6	0.9	0.6	<0.5	<0.5	0.6	<0.5	<0.5
Tb	ppm	0.1	0.6	0.5	0.6	0.5	0.6	0.5	0.5	0.5	0.5	0.4	0.5	0.5
Th	ppm	0.2	6.9	6.6	7.3	6.9	8.4	8.0	5.4	8.8	8.3	7.6	8.8	8.2
Tl	ppm	0.5	0.6	0.9	0.9	1.2	1.6	1.7	<0.5	<0.5	<0.5	<0.5	<0.5	<0.5
Tm	ppm	0.1	0.2	0.2	0.2	0.2	0.2	0.2	0.2	0.1	0.1	0.1	0.1	0.1
U	Ppm	0.1	1.4	1.4	1.5	1.5	1.9	1.7	1.1	1.8	1.7	1.5	1.7	1.7
W	Ppm	1	4	1	<1	1	2	1	2	2	2	2	1	2
Yb	Ppm	0.1	1.2	1.0	1.1	1.1	1.2	1.0	1.1	0.8	0.9	0.9	0.9	0.9

3.5 X-ray diffraction

X-ray diffraction shows a dominance of plagioclase in its anorthite variety and volcanic glass (Figure 3).

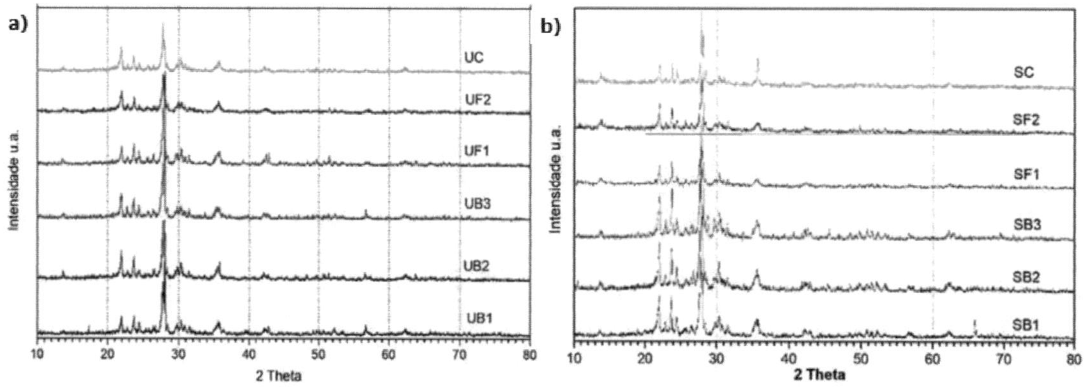

Figure 3. X-ray diffraction of studied volcanic ashes dispersed by Ubinas (a) and Sabancaya (b) volcanoes.

Ash characterization based on SEM and XRD analyses shows that glass is the main constituent of Ubinas and Sabancaya ashes, which appears to be accompanied by plagioclase, mainly andesine, quartz, and cristobalite [5]. The collected ashes present several mineral components, such as plagioclase, olivine, biotite, lithics and volcanic glass, as well as hydro-thermalized and oxidized lithics.

For Ubinas, other samples collected in March and April 2006 in medial areas (between 5 and 8 km distance from the volcano) contained a dominant 50-100 μm grain-size fraction, and less abundant fine ash. The large particles were mostly shards of andesitic composition (56–57 wt.% SiO_2) [5]. Samples included in the present research contains particles between 5–600 μm grain size, a dominant size over 100 um, and are mostly shards of andesitic composition for Ubinas (58- 63 wt.% SiO_2) and andesitic and dacitic for Sabancaya (57–67 wt.% SiO_2) [3]. According to the authors, volcanic glass and pyroxene are the prevalent components. Samaniego (2016) found also amphibole and biotite (similar to Sabancaya and Ubinas in Figure 3) [17].

Compositions range from andesite to dacite, and also exhibit the high potassium contents typical of the stratovolcanoes of the Central Volcanic Zone of the Andes and southern Peru (Figure 4).

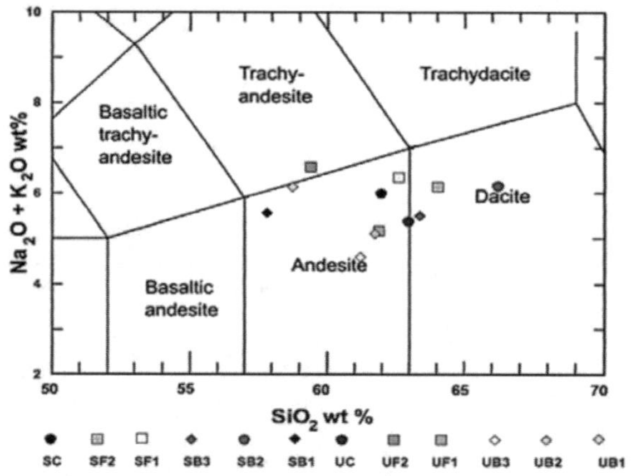

Figure 4. Classification of ash based on the alkali diagram vs. silica.

The variation in composition suggests that these grains may share a common magmatic source. The absence of silicification indicates the prolonged exposure to the hydrothermal system in some cases; samples from the base and hillsides of the volcano are different to those from craters. The distance among samples as well as the altitude can play a role in those differences due to different hydrothermal conditions.

Volcanic ashes have been used for alkaline activation or as fillers in civil engineering [18]. They can be considered as pozzolans due to their high silica, alumina and iron oxide amounts. In order to manufacture geopolymers, particle size distribution, specific surface, chemical and mineralogical composition, available silica and alumina, among other factors, have to be checked to assure alkaline activation of aluminosilicates [19]. Other parameters (curing temperature, age of geopolymers, activating alkaline solution and the addition of aggregates) affect performance of manufactured materials. Chemical and mineralogical compositions of volcanic ashes are variable but can be used to manufacture geopolymers [20] [21].

The importance of natural materials (as volcanic ash) in construction generates profit for circular economy as well as environmental benefits, as emitted CO_2 is strongly reduced.

4. Conclusions

Based on specific gravity test results on volcanic ashes in Ubinas and Sabancaya, they can be classified in light material category. The chemical composition of the ashes is mainly SiO_2, Al_2O_3, Na_2O and CaO with a high content of silica rangeing from 57.78 to 66.14%, while alumina ranges from 13.81-17.56%. This can be a reference for further research on the utilization of materials of volcanic ash as a pozzolanic material in soil stabilization. The chemical composition and mineralogy also showed that ashes have an andesitic and dacitic origin and formation. The narrow variation in composition suggests that the studied ash may share a similar source of magma characteristic of volcanoes in the Central Volcanic Zone from Peru.

Acknowledgments

The authors acknowledge the Carlos III University of Madrid and INGEMMET for their collaboration and Dr. Carol Paredes for her unconditional support.

References

[1] Smithsonian Institution 2000. *Bulletin of the Global Volcanism Network* CrossRef.

[2] Thouret JC, Rivera M, Wörner G, Gerbe MC, Finizola A, Fornari M and Gonzales K 2005 Ubinas: The evolution of the historically most active volcano in southern Peru. *Bulletin of Volcanology.* **67** 557–89 CrossRef.

[3] Rivera M, Thouret J-C, Mariño J, Berolatti R and Fuentes J 2010 Characteristics and management of the 2006–2008 volcanic crisis at the Ubinas volcano (Peru). *Journal of Volcanology and Geothermal Research.* **198** 19–34 CrossRef.

[4] Ticona PE 2014 Diagnostico de la calidad del aire en la zona circundante al volcán San Pedro del distrito de Ubinas. *Dirección Regional de Salud Moquegua* CrossRef.

[5] Rivera M, Thouret JC, Samaniego P and Le Pennec JL 2014 The 2006-2009 activity of the Ubinas volcano (Peru): Petrology of the 2006 eruptive products and insights into genesis of andesite magmas, magma recharge and plumbing system. *Journal of Volcanology and Geothermal Research.* **270** 122–41 CrossRef.

[6] Instituto Geográfico Nacional del Perú (IGN) 2017 *Carta Geológica Nacional* 1:100000. Hoja 32-s (2539) Serie J631 CrossRef.

[7] Zamácola y Jáuregui JD 1951 *Los terremotos en Arequipa:1582-1868.* Barriga VM La Colmena Arequipa 332.

[8] Thouret JC, Guillande R, Huaman D, Gourgaud A, Salas G and Chorowicz J 1994 L'Activite actuelle du Nevado Sabancaya (sud Perou); reconnaissance geologique et satellitaire, evaluation et cartographie des menaces volcaniques. *Bulletin de la Société Géologique de France.* **165** 49-63

CrossRef.
[9] Observatorio Vulcanológico de INGEMMET 2016 Informe anual 2016: Monitoreo de la actividad del volcán Sabancaya 1-29 CrossRef.
[10] Ruggieri F, Fernandez-Turiel JL, Saavedra J, Gimeno D, Polanco E, Amigo A, Galindo G and Caselli A 2012 Contribution of volcanic ashes to the regional geochemical balance: The 2008 eruption of Chaitén volcano, Southern Chile. *Science of the Total Environment.* **425** 75–88 CrossRef.
[11] Tupayachy-Quispe D, Almirón J, Apaza F, Churata R, Paredes E, Torres-Carrasco M and Bautista A 2020 Peruvian volcanic ashes as new alternative material in geopolymer preparation: influence of dissolution concentration and wear resistance. *18th LACCEI International Multi-Conference for Engineering, Education, and Technology* **5** 1-7 CrossRef.
[12] Almirón J, Roudet F and Duquesne S 2019 Influence of volcanic ash, rice husk ash, and solid residue of catalytic pyrolysis on the flame-retardant properties of polypropylene composites *Journal of Fire Sciences.* **37** 434–51 CrossRef.
[13] Devi Oktaviana L, Ahmad R and Kabul Basah S 2016 Chemical characteristics of volcanic ash in Indonesia for soil stabilization: Morphology and mineral content. *International Journal of GEOMATE.* **11** 2606-10.
[14] Observatorio Vulcanológico del Sur – OVS 2017 Informe anual 2017 - Monitoreo de la actividad del volcán Sabancaya 1-29 CrossRef.
[15] Heiken G. 1972 Morphology and petrography of volcanic ashes. *Geological Society of America Bulletin.* **83** 1961-88 CrossRef.
[16] Wohletz KH 1983 Mechanisms of hydrovolcanic pyroclast formation: Grain-size, scanning electron microscopy, and experimental studies. *Journal of Volcanology and Geothermal Research.* **17** 31–63 CrossRef.
[17] Samaniego, P. (2016). The eruptive chronology of the Ampato–Sabancaya volcanic complex (Southern Peru). *Journal of Volcanology and Geothermal Research.* **323** 110-128 CrossRef.
[18] Kamseu E, Leonelli C, Perera DS, Melo UC and Lemougna PN 2009 Investigation of volcanic ash based geopolymers as potential building materials. *International Ceramic Review.* **58** 136–140 CrossRef.
[19] De Silva P, Sagoe-Crenstil K and Sirivivatnanon V 2007 Kinetics of geopolymerization: role of Al_2O_3 and SiO_2. *Cement and Concrete Research.* **37** 512–518 CrossRef.
[20] Tchakoute HK, Elimbi A, Yanne E and Djangang CN 2013 Utilization of volcanic ashes for the production of geopolymers cured at ambient temperature. *Cement and Concrete Composites.* **38** 75–81 CrossRef.
[21] Calderón N, Vargas M, Almirón J, Bautista A, Velasco F and Tupayachy-Quispe D 2021 Influence of the activating solution and aggregates in the physical and mechanical properties of volcanic ash based geopolymer mortars. *IOP Conference Series Materials Science and Engineering.* **1054** 1–9 CrossRef.

Influence of the activating solution on the mechanical properties of compacted volcanic ash based geopolymers

Rossibel Churata[1], Jonathan Almirón[2], Francine Roudet[3], Asunción Bautista[4], Manuel Torres-Carrasco[4], Danny Tupayachy-Quispe[5]

[1]Escuela Profesional de Ingeniería de Materiales, Facultad de Ingeniería de Procesos, Universidad Nacional de San Agustín de Arequipa. Av. Independencia 400, Cercado – Arequipa, Perú.
[2]Escuela Profesional de Ingeniería Ambiental, Facultad de Ingeniería de Procesos, Universidad Nacional de San Agustín de Arequipa. Av. Independencia 400, Cercado – Arequipa, Perú.
[3]Laboratoire de Génie Civil et géo-Environnement, Université de Lille, Villeneuve d'Ascq 59650, France.
[4]Materials Science and Engineering Department, IAAB, Universidad Carlos III de Madrid. Avda. Universidad Nº 30. 28911 - Leganés, Madrid, Spain.
[5]Laboratorio de Ciencia de los Materiales, Universidad Católica de Santa María, Calle Samuel Velarde 320, Umacollo, Arequipa, Perú.

E-mail: dtupayachy@ucsm.edu.pe

Abstract. This work studies the mechanical properties of volcanic ash based geopolymer mortars. The geopolymer mortars were made with volcanic ashes from Ubinas volcano using mixtures of Na2SiO3 and 12 M NaOH as alkaline activator. Solutions with different concentrations of silicate (0-25 % by wt.) were used. The geopolymer mortars were compacted at 15 MPa and cured at 80°C for three different periods of time. The results show the influence of the concentration of Na2SiO3 in the compression strength and wear resistance, raising highest compression strengths for 20-25% of Na2SiO3 while the wear resistance was higher at concentrations over 20% of Na2SiO3. In this optimal ratio, for 20% Na2SiO3, a compression strength up to 52.3 MPa was recorded at 7 days. Considering the overall performance analysis, the synthesis of geopolymers with volcanic ash promises a solution for sustainable mortar production and natural waste disposal.

1. Introduction

The volcanic ashes are vitreous pyroclastic materials produced by an eruptive volcanic action [1]. The volcanic ashes are usually grey and dark grey. They can also have a huge impact on the farmland near the volcano because during the raining season tephra and lahars falls down [2]. The composition of volcanic ashes makes them potentially attractive for the construction industry to produce geopolymers [3] [4]. Additionally, studies of volcanic ashes as fire retardant additive load on polymers has been started [5].

The geopolymer is an inorganic polymer based on alumina and silicates which is synthetized by silicon (Si) and aluminum (Al) material, that can be from geological origin [6]. There are some variables that influence geopolymerization process, such as molar concentrations of the alkali and sodium silicate, curing temperature, liquid:solid ratio, etc. A study of the effect of Na_2SiO_3 at 24 h of curing has reported that high concentrations of Na_2SiO_3 have influence on the specimens, while those high concentrations of Na_2SiO_3 at lower temperatures of curing result in lower compression strengths [7].

However, there are not many studies about the use of volcanic ash for the production of geopolymers and the variables that influences the synthesis. As a result, this work studies the influence of different concentrations of Na_2SiO_3 as alkali activator on the mechanical properties of volcanic ash based geopolymer mortars

2. Materials and methods

2.1. Materials

The volcanic ashes used in this work were obtained from the Ubinas volcano located in Moquegua (Peru). The volcanic ashes were ground in a planetary ball mill Fritsch 6S to obtain powder with a particle size namely under 75 μm. Figure 1 shows the particle size distribution of the volcanic ashes which was determined by a laser light scattering particle analyzer Malver MASTERSIZER 2000. The real particle size distribution ranges between 0.3 and 100 μm, with diameters of d(50)= 25.88 μm and d(90)= 57.42 μm. The density of the particles size distribution (blue curve in Figure 1) shows that powder of volcanic ash has a bimodal distribution. The first mode ranges from 0.3 to 12 μm and second mode ranges from 12 to 100 μm, being the median particle size 27.39 μm. The grounded volcanic ash has a specific gravity of 2.62 $g \cdot cm^{-3}$ and a specific surface of 2.29 $m^2 \cdot g^{-1}$.

Additionally, the chemical composition of the volcanic ashes was determined by inductive plasma (ICP) mass spectrometry. Table 1 shows the presence of the main oxides such as SiO_2, Al_2O_3, and Fe_2O_3; with a molar SiO_2/Al_2O_3 ratio of 3.96. This composition makes the volcanic ashes appropriated as a raw material for the synthesis of geopolymers according to the basic bibliography for synthesis of geopolymers [8].

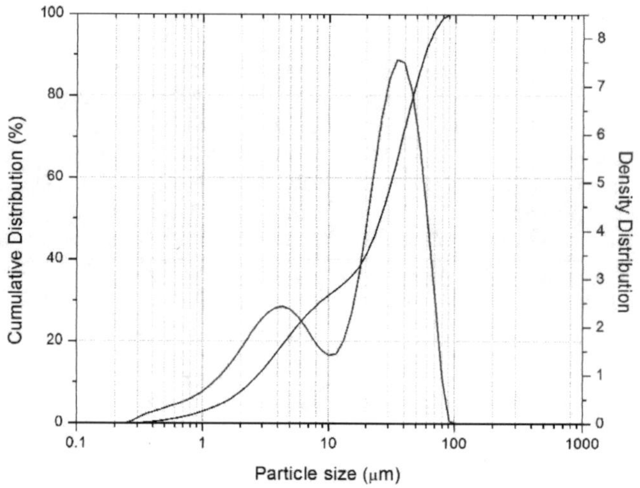

Figure 1. Particle size distribution of the milled volcanic ashes of Ubinas.

Table 1. Chemical composition (%) of Ubinas volcanic ash.

Oxide	SiO_2	Al_2O_3	Fe_2O_3	CaO	MgO	K_2O	Na_2O	P_2O_5	MnO	TiO_2	LOI*
%	58.70	14.80	7.69	5.80	4.17	2.63	3.52	0.37	0.10	1.14	0.87

LOI: Loss on Ignition

The mineralogical composition of volcanic ashes was determined by a X ray diffraction (XRD). Figure 2 shows that anorthite $(Ca,Na)(SiAl)_4O_8$ is the main mineralogical component in volcanic ashes. In addition, the morphology of particles was determined by scanning electron microscopy (SEM). Figure 3 shows SEM images of the volcanic ashes particles at 5000x magnification; as it is shown, the morphology is irregular and varies on its shape and size. Most of the particles has an angular shape due to the grounding.

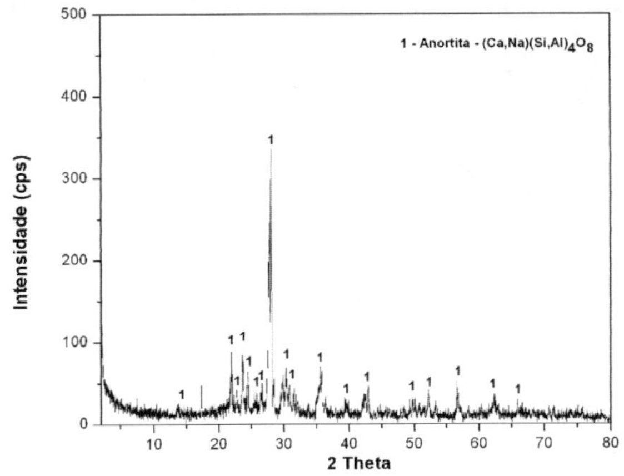

Figure 2. X-ray diffraction analysis of Ubinas volcanic ash

Figure 3. SEM micrograph of Ubinas volcanic ash. 5000x

2.2. Geopolymer synthesis

The geopolymers were obtained by an alkali activation of the volcanic ashes, using a liquid/solid ratio of 0.2, with an alkali activator solution [3]. It is reported that activating solutions with both sodium silicate and sodium hydroxide increase the geopolymerization process and improve the strength of mortars [9]. The concentration of sodium hydroxide and the proportion of Na_2SiO_3/NaOH influence the properties of fresh and strengthen geopolymer mortars [10].

The alkali solution was then based on sodium silicate (Na_2SiO_3) and sodium hydroxide (NaOH), in a concentration of 12M NaOH with a mass percentage Na_2SiO_3 of 0, 5, 10, 15, 20 and 25%. Each solution was denominated as G0, G5, G10, G15, G20 and G25 respectively. The geopolymer paste was mixed for 5 min. Then, it was poured into cylindrical molds of 28 mm diameter, followed by a compression at 15 MPa; this action reduces the amount of alkali solution to be used in the geopolymerization concerning the cost. The specimens were immediately removed from the mold, covered with plastic film and cured at 80 °C for 48 h in a Memmert brand forced convection oven.

2.3. Characterization techniques

The bulk density of dried mortars was determined according to the ASTM C140, which calculates the density after 28 days of curing. The water adsorption was determined by the ASTM C-642 and the compression strength was studied at 7, 14 and 28 days of curing according to the ASTM C39 using an automatic compression testing machine (ELE International, England).

The wear resistance was determined in a Microtest pin-on-disc machine according to the ASTM G99-03 using a stainless steel ball (AISI 316) of 6 mm diameter, applying a load of 15 N at 60 rpm.

3. Results and discussion

Figure 4 shows the density of geopolymer mortars and the water adsorption at different mass concentrations of Na_2SiO_3 (0, 5, 10, 15, 20 and 25%). The bulk density of the cured/hardened geopolymer samples ranges from 1.78 to 2.06 g·cm^{-3}. The results indicate that the density of volcanic ashes based geopolymer pastes increases from 5 to 25% with increasing Na_2SiO_3. More than the fact that Na_2SiO_3 - NaOH solutions have higher density than water which makes the activation solution denser, the Si present in the solution enhances the polymerization process, thus increasing the density of the gel formed (probably N-A-S-H gel) in the mortar [11].

Increasing the concentration of sodium silicate has a small but noticeable effect on water adsorption, because water adsorption decreases with the increasing of sodium silicate. Water adsorption is clearly related to porosity in the samples, which is opposite to density [12]. The increases of density shown in Figure 1 are in accordance with the reduction of water absorption. This behavior was observed in [12], where the increase of sodium silicate reduces the water adsorption until 17, 13 and 15%.

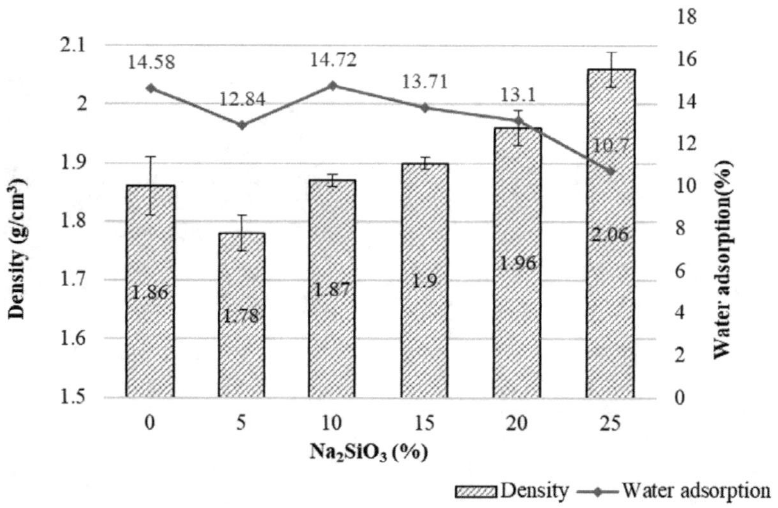

Figure 4. Bulk density of geopolymers (main y-axis) and water adsorption (secondary y-axis) related to the amount of sodium silicate in the activating solution.

Figure 5 shows the compression strength at 7, 14 and 28 days of curing of the geopolymers with different concentrations of Na_2SiO_3 and NaOH. It is shown that the compression strength increased at 28 days of curing with the increasing mass of Na_2SO_3 with values of 32, 33, 41, 44 and 51 MPa for samples G0, G5, G10, G15, G20 and G25 respectively. Therefore, increasing the amount of sodium silicate the compression strength will increase. For samples G0 and G5 (which have less viscous activation solution), the compression strength is higher at 7 and 14 days of curing which can be attributed to the concentration of Na_2SiO_3 which increase the viscosity in the activation solution. The viscosity of the solution influences the wettability, therefore the geopolymer samples need more quantity of solution to achieve a complete geopolymerization and, as a consequence, the compression strength can be reduced. The addition of soluble silicates affects the polymerization of the composite, the setting time and viscosity; furthermore, it is necessary to consider the concentration of the alkali in order to have the enough concentration for a balance in the substitution of the silicon tetrahedron by the aluminum and avoid the lixiviation and its reaction with CO_2. The CO_2 will result in a formation of carbonates or hydrogen carbonates in an efflorescence form or precipitations decreasing the compression strength [13].

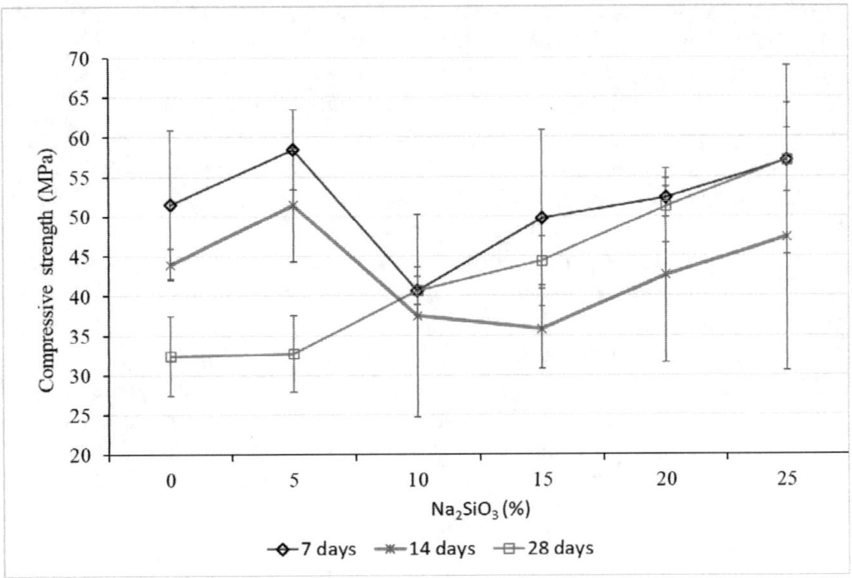

Figure 5. Influence of the increasing Na_2SiO_3 in the activation solution over the compression strength of geopolymers at 7, 14 and 28 days of curing.

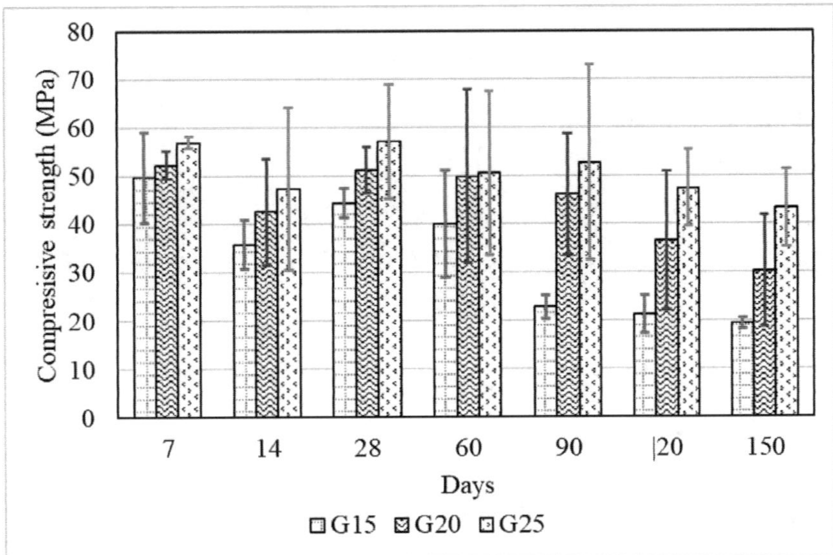

Figure 6. Influence of the curing time from 7 to 150 days on the compression strength

In order to study the effects of the geopolymers curing time on the compression strength, compression tests were carried out at 60, 90, 120 and 150 curing days. Under the 28 days of curing the effect of Na_2SiO_3 concentration on the compression strength was similar to the studied in figure 5 [7]. Figure 6 shows that the compression strength for sample G15 decrease followed by samples G20 and G25. The maximum value for sample G15 was 49.74 MPa at 7 days of curing, for sample G20 was

52.35 MPa at 7 days of curing and for sample G25 was 57.1 MPa at 28 days of curing. The chemical reaction of the geopolymer in the geopolymerization process is fast, the geopolymer achieve the 70% of its strength in the first 3-4 h of curing. It is expected that the compression strength will not vary with time due to the 24 hours of accelerated thermal curing employed. However, this work revels a decreasing compression strength starting the 28 days of curing which can be attributed to an excess of NaOH that is shown as an efflorescence form and can be cracking the samples. Additionally, the durability of the active alkali conglomerates depends on the microstructure of the reaction products on the system which rely on the prime material, nature and concentrations of the activator and curing time [13].

The ASTM G99-03 standard was used to calculate the wear rate (Figure 7). It is possible to determine the lowest wear rate which means a highest durability in sample G20 as it is shown. The combination of the activator geopolymer in sample G20 has shown good results on the compression strength test, so it can be inferred that the combination of 20% de Na_2SiO_3 and 12 M de NaOH promotes better reaction and hence better mechanical properties; on the other side, the wear strength is more than two orders of magnitude lower than a sample of cement (CEM IV) ($6{,}62 \times 10^{-5}$ $mm^3/N \cdot mm$) after 28 days of curing. For sample G0 (12 M NaOH activating solution), the wear strength compared to the same conventional cement is also lower. However, for samples G5, G10, G15 and G25 there is a decrease on the wear strength compared to other tested geopolymers.

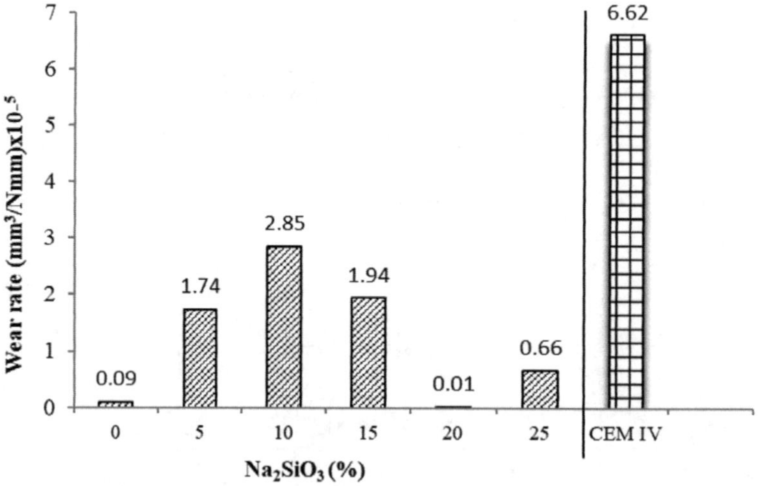

Figure 7. Wear behavior in the different materials tested (CEM IV and geopolymeric pastes activated with 12 NaOH at different amounts of Na_2SiO_3).

4. Conclusions

This work has studied the behavior of the volcanic ashes based geopolymer mortar from the Ubinas volcano located in Peru, which had different concentrations of the Na_2SiO_3 in the activation solution, leading to the following conclusions:

- The amount of Na_2SiO_3 in the activator alkali is a critic parameter related to the compression strength and wear resistance of geopolymer samples. Another parameter such as the curing time of samples has affect the compression strength. The best composition of alkali activation solution was obtained for 20 % Na_2SiO_3 with 12 M NaOH.
- The 12 M NaOH as an alkali activator along with different concentrations of Na_2SiO_3 generates a high mechanical performance at short periods. The composition of the alkali solutions plays an essential role on the dissolution of silicon and aluminum from volcanic ashes. The wear

IOP Conf. Series: Materials Science and Engineering 1150 (2021) 012011 doi:10.1088/1757-899X/1150/1/012011

durability of the volcanic ashes alkali activated with different concentrations suppose an increase of 99.8% on the wear strength compared to a sample of cement.

Acknowledgements
We wish to express our gratitude to Luis Cesar Perez Huamaní. This study was funded by the Ministerio de Ciencia, Innovación y Universidades of Spain through project RTI2018-096428-B-I00 and through regional project HORATSO-UC3M-CM.

Article I.

References

[1] Lemougna PN, Melo UF, Delplancke MP, Rahier H. Influence of the chemical and mineralogical composition on the reactivity of volcanic ashes during alkali activation. Ceramics International. 2014; 40: p. 811– 820.
[2] Rivera M, Thouret JC, Samaniego P, Le Pennec JL. The 2006– 2009 activity of the Ubinas volcano (Peru): Petrology of the 2006 eruptive products and insights into genesis of andesite magmas, magma recharge and plumbing system. Journal of Volcanology and Geothermal Research. 2014; 270: p. 122–141.
[3] Tupayachy-Quispe D, Almirón J, Apaza F, Churata R, Paredes E, Torres-Carrasco M, et al. Peruvian volcanic ashes as new alternative material in geopolymer preparation: influence of dissolution concentration and wear resistance. In 18th LACCEI International Multi-Conference for Engineering, Education, and Technology: "Engineering, Integration, and Alliances for a Sustainable Development"; 2020; Buenos Aires.
[4] Calderón N, Vargas M, Almirón J, Bautista A, Velasco F, Tupayachy-Quispe D. Influence of the Activating Solution and Aggregates in the Physical and Mechanical Properties of Volcanic Ash Based Geopolymer Mortars. In The 4th International Conference on Building Materials and Materials Engineering; 2020; España.
[5] Almirón J, Roudet F, Duquesne S. Influence of volcanic ash, rice husk ash, and solid residue of catalytic pyrolysis on the flame-retardant properties of polypropylene composites. Journal of Fire Sciences. 2019; 37(4-6): p. 434-451.
[6] Davidovits J. Geopolymer chemistry and application: France: Geopolymers Institute; 2008.
[7] Mijarsh MJA, Megat Johari MA, Ahmad ZA. Effect of delay time and Na2SiO3 concentrations on compressive strengthd evelopment of geopolymer mortar synthesized from TPOFA. Construction and Building Materials. 2015; 86(1): p. 64.74.
[8] Bondar D, Lynsdale CJ, Milestone NB, Hassani N, Ramezanianpour AA. Effect of type, form, and dosage of activators on strength ofalkali-activated natural pozzolans. Cement and Concrete Composites. 2011; 33: p. 251–260.
[9] Nematollahi B, Sanjayan J. Effect of different superplasticizers and activator combinations on workability and strength of fly ash based geopolymer. Materials & Design. 2014; 57: p. 667-672.
[10] Hanjitsuwan S, Hunpratub S, Thongbai P, Maensiri S, Sata V. Effects of NaOH concentrations on physical and electrical properties of high calcium fly ash geopolymer paste. Cement and Concrete Composites. 2014; 45: p. 9-14.
[11] Çelikten S. Mechanical and microstructural properties of waste andesite dust-based geopolymer mortars. Advanced Powder Technology. 2020.
[12] Mohseni E. Assessment of Na2SiO3 to NaOH ratio impact on the performance of polypropylene fiber-reinforced geopolymer composites. Construction and Building Materials. 2018; 186: p. 904–911.
[13] Škvára F, Šmilauer V, Hlaváček P, Kopecký L, Cílová Z. A weak alkali bond in (N,K)-A-S-H gels: Evidence from leaching and modeling. Ceramics-Silikáty. 2012; 56(4): p. 374–382.

IOP Conf. Series: Materials Science and Engineering 1150 (2021) 012012 doi:10.1088/1757-899X/1150/1/012012

A combination of water hyacinth (*Eichhornia crassipes)* and coffee grounds as a growth substrate for use in arid areas

Quang Huu Le[1*], Huong Thi Nguyen[1], Au Hai Nguyen[2]

[1]Center of Water Management and Climate Change, Institute for Environment and Resources, Vietnam National University – HoChiMinh City, Vietnam
[2]Institute for Environment and Resources, Vietnam National University – HoChiMinh City, Vietnam

E-mail: lequang@wacc.edu.vn

Abstract: There is a strict requirement to reduce the environmental impact of the water hyacinth and coffee grounds and an urgency to find solutions to boost farming in arid regions. In this experiment, growth substrates containing different ratios of water hyacinth and coffee grounds were used to support potted flower growth to evaluate their effectiveness as soil replacements. Two substrates were prepared by the combination of water hyacinth and coffee grounds ratios of 50:50 and 70:30, and then their physicochemical properties were measured. The growth, biomass production, and micro and macroscopic nutrient uptake of cockscomb flower (*Celosia argentea* var. *cristata*) were also determined. The findings suggest that the addition of organic raw materials improved phytonutrients and increased substrate pH and electrical conductivity (EC) of growing medium. The experimental growth substrates did not negatively affect plant growth. Overall, the results have shown that the application of water hyacinth and coffee grounds in growth substrates reduces their negative impact on the environment and increases the amount of arable land in arid, water-scarce areas with poor soil quality.

1. Introduction

Currently, in arid and water-deficient areas, there is a need for growth substrates that retain water and provide nutrients for plants. If natural, readily available materials could be used as growth substrates, then these pressing problems could be solved in a way that protects the environment. Growth substrates are a mix of different ingredients, such as peat, coir, wood fiber, and bark [1]. In 2012, Mohanty and colleagues studied the growth of *Cymbidium masterii* orchid. Orchids were grown into a substrate consisting of crushed sterile charcoal and bricks (1:1) and worm manure and found that the survival rate increased up to 88% [2]. In 2017, Srinivasulu and colleagues tested different growth substrates for the *Dendrobium* variety [3] and found the positive effects of added materials.

Water hyacinth *(Eichhornia crassipes)* is probably the worst aquatic invasion plant in most tropical countries [4]. It proliferates and forms a dense carpet [5], obstructs waterways, destroys aquatic wildlife, creates ideal conditions for disease and vector transmission, and more [6]. However, several applications of this plant can be employed such as use as animal feed, in water purification, as a biogas fuel, as fertilizer, and in paper production [7]. With the current rapid proliferation of water hyacinth, researchers have been searching for an approach to solve the problem of dense growth in canals, rivers, and other waterways that can bring environmental benefits and economic value.

Coffee grounds are by-products during the production of instant coffee. This material has high nutrient and organic content. Studies have been conducted to examine the use of coffee grounds as artificial soil, which can also contribute to limiting waste emissions into the environment. The substrate is used to support plant growth by providing nutrients as well as air and water to the roots, which allows the roots to develop as much as possible.

2. Methods

2.1. Preparation of sample.

The following compounds were mixed well and gathered into a heap 1–1.5 m high: water hyacinth root (about 500–700 kg), coffee grounds (300–500 kg), phosphate fertilizer 16% (2 kg), lime (1 kg), *Trichoderma* sp. (50 g), and cotton powder as binder (1 kg). The heap was watered and compacted, then sealed in a nylon tarpaulin to retain moisture and watered weekly. After three weeks, the canvas was opened and the heap turned, then sealed again. After 1.5 months of incubation, the material was used directly or mixed with binder to create other media such as tablets for growing orchids. Seven days- germinated cockscomb was chosen for the experimental.

2.2. Method of experimental design.

The efficiency of the experimental growth substrates was assessed as follows. First, three pots were prepared with 5 kg of soil each, and the soil was taken from arid locations. Then, the experimental growth substrate was added to two of the pots (P1 = 500 kg water hyacinth root + 500 kg coffee grounds, P2 = 700 kg water hyacinth root + 300 kg coffee grounds). The third pot was used as a control (P0). Cockscomb was planted to each pot with the same number and size of trees per pot. In all experiments, the pots were watered twice a day. The experiment was repeated three times. After two months, the cockscomb height, the number of flowers/plants, and the diameter of the main flower were measured, and the nutrient composition of the soil and the growth substrates were analyzed to determine the efficiency of these materials.

2.3. Sample analysis

For the original soil and growth substrates, nutrient content and physical properties were analyzed. The nutrient composition of the soil was analyzed using the classical physiological method. Data processing was performed using Excel and Statgraphics Centurion XV software (Statgraphics Technologies, USA).

3. Results and discussion

Physical properties are an important element of the soil as well as any substitute material. They determine the quality of the growth substrate and the availability of nutrients for plants [8]. Table 1 shows that the soil and the replacement materials had pH indices suitable for plant growth. Specifically, pH ranged from 5.83–6.50, within the range recommended for soil that supports plant growth. Particularly in the two growth substrates, because they were treated with phosphorus and lime beforehand, the pH was raised from 0.53–1.8. Regarding EC, the substrates show better EC than the soil. A higher EC indicates higher substrate quality and better metabolism between the substrate and the crop. In addition, the moisture in the growth substrates was higher than that in the control, which was the basis for reducing the amount of irrigation (2 days/time compared to the usual 2 times). This provided evidence of the water-holding capacity and water-saving ability of the experimental substrates.

Table 1. Physical and physicochemical characteristics of the growth substrates

Treatment	pH_{KCl}	EC(dS/m)	Moisture content (%)
P0	5.83	0.48	13.2
P1	6.36	2.15	28.1

P2	6.50	2.32	24.3

In Table 2, the obtained results show that the organic composition of the two growth substrates was about 10 times higher than that of the control. This proves that when the organic content of the growth substrates' raw materials is high, it is beneficial for the growth of plants, increasing the ability to retain water in the soil. Similarly, the major macronutrients (N, P, and K) required for plant growth, the content of all three elements was 10–20 times higher than the content in the control.

Table 2. Chemical characteristics of the growth substrates

Treatment	OC (%)	N(%)	P_2O_5(%)	K_2O (%)
P0	2.18	0.02	0.15	0.68
P1	18.2	1.42	1.17	1.12
P2	21.4	1.78	2.06	1.46

Table 3 presents the growth and development factors for the plant development in different growth substrates. The data show that the growth rate of the plants in the experimental growth substrate was higher than those in the control substrate. Also, the tallest plant reached 32.7 cm after one month in the experimental substrate, whereas the tallest in the control substrates reached 29.7 cm. This difference is statistically significant. The results in Table 3 also show that the average number of flowers/plants produced in the control was lower than those produced by the two experimental treatments. It is also worth noting that the size and area of flowers grown in the experimental substrate tended to be larger than those grown in the control, but there was no statistically significant difference.

Table 3. The growth measurements of the plants grown in the three different substrates

Treatment	Height (cm)			The number of flowers/plant
	10 days	20 days	30 days	
P0	13.40^{a}	22.03	27.90^{a}	11
P1	15.43^{ab}	25.77	31.83^{b}	13
P2	15.80^{b}	26.03	32.70^{b}	12
CV (%)	9.6	12.5	8.7	20.1
LSD	2.1	NS	3.5	NS

Note: LSD: least significant difference test, NS: none significant, CV: Coefficient of Variation.

Figure 1 presents the blossoming and flowering times of the plants that were grown in the experimental growth substrates. These results confirm that the experimental growth substrates provided a better nutritional composition and physical factors than the control substrate, which helped the flowering plants to grow well. Flowers appeared earlier and lasted longer, averaging nearly 30 days from the date of bloom.

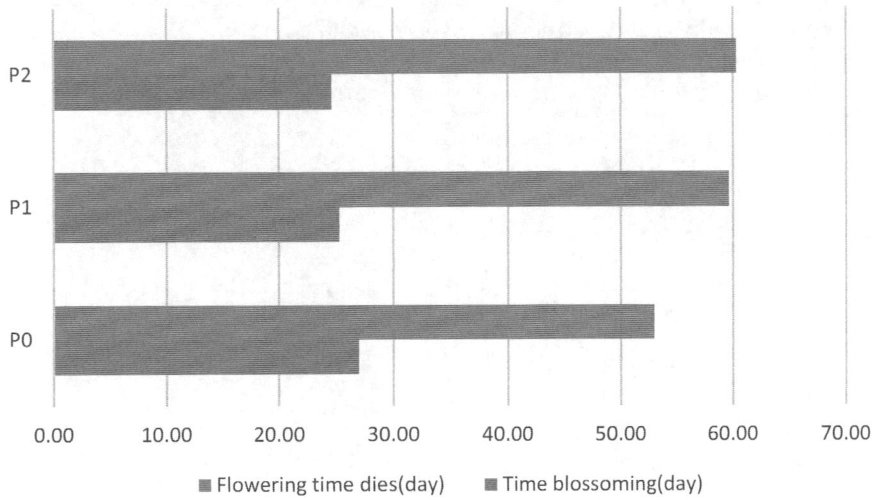

Figure 1. The time blossoming and the flowering time (day).

4. Conclusions

The results show that the replacement of arid soil with a growth substrate containing coffee grounds and water hyacinth can increase water and nutrient availability of plant growing media, leading to better plant growth. Plant growth is improved due to the growth substrates increasing the supply of macronutrients, organic matter, and water to plants. In our experiments, there was no significant difference in the cockscomb growth between the two experimental substrates, which had different ratios of ingredients. The results showed that the addition of water hyacinth and coffee grounds to soil significantly reduces their environmental impact and is also a way to grow plants in dry, dehydrated areas.

Acknowledgments

This work was financed by Vietnam National University – Ho Chi Minh City, Vietnam (VNU) through project number "C2020-24-08"

References

[1] Grunert O, Reheul D, Van Labeke, M-C, Perneel M, Hernandez-sanabria E., Vlaeminck SE and Boon N 2016. Growing media constituents determine the microbial nitrogen conversions in organic growing media for horticulture. *Microbial Biotechnology* 9 (3) 389-399.
[2] Mohanty P, Paul S, Das MC, Kumaria S, Tandon P 2012. A simple and efficient protocol for the mass propagation of Cymbidium mastersii: an ornamental orchid of Northeast India. *AoB PLANTS,* pls023-pls023.
[3] Srinivasulu G, Harshavardhan M, Chandan K 2017. Comparative effect of different potting media on vegetative and reproductive growth of Dendrobium orchid var Sonia-17. *Environment and Ecology,* 35, 1252-1255.
[4] Boyd CE 1968. Fresh-water plants: a potential source of protein. *Economic Botany,* 22, 359-368.
[5] Gopal B, Goel U 1993. Competition and allelopathy in aquatic plant communities. *The Botanical Review,* 59, 155-210.
[6] Bolch E, Santos MJ, Ade C, Khana S, Basinger NT, Reader MO, Hestir EL 2012. Remote Sensing of Plant Biodiversity. In: Cavender-Bares J., Gamon J.A., Townsend P.A. (eds) Remote Sensing of Plant Biodiversity. Springer, Cham. https://doi.org/10.1007/978-3-030-33157-3_12.

[7] Lindsey K, Hirt HM 1999. Use water hyacinth! a practical handbook of uses for the water hyacinth from across the world. Physical description, Environmental Science.
[8] Raviv M 2005. Production of high-quality composts for horticultural purposes: a mini review. Production of high-quality composts for horticultural purposes: a mini-review. HortTechnology 15(1): 52-57.

A Bayesian Inference Driven Computational Framework Applied for the Improvement of the Eurocode 2 creep model

H Daou[1,2], W Raphael[1] and F Geara[1]

[1] Saint-Joseph University of Beirut, Lebanon

E-mail: hikmat.daou@net.usj.edu.lb

Abstract. Concrete creep, defined as the deformation of concrete under sustained load, can cause cracks in tensile members, redistribution of stresses over time in composite structures, loss of prestressing force in prestressed concrete elements, and excessive long-term deflection of structural members. The concrete creep coefficient is an important input in many calculations and analyses of reinforced concrete structures. Currently, the concrete creep coefficient has been predicted by many models such as the Eurocode 2 model. This study aims to improve the prediction of the Eurocode 2 creep coefficient model at long-term by implementing a correction coefficient into the model. The Northwestern University database is used and the correction coefficient is calculated using Bayesian inference. The accuracy and efficient of the proposed improvement and modification are demonstrated through statistical indicators.

1. Introduction

The creep of concrete is of great practical importance to structural engineers; therefore, it has been studied by many researchers [1–4]. Creep, defined as a time-dependent behavior of concrete, can affect structural behavior by violating service limit states, losing prestressing forces, or redistribution stress [5–9]. Therefore, designers must accurately predict creep strains using precise methods.

The concrete creep coefficient is an important input in many analyses and calculations of reinforced concrete structures, and it can be predicted using the Eurocode 2 model (EC2) [10], which is one of the most widely used models for predicting creep. Based on a large experimental database, the EC2 shrinkage model has been updated [11], and correction coefficients have been also proposed for the Eurocode 2 creep model [12] but the calculation of these correction coefficients are limited for a specific condition of initial time loading, relative humidity, or compressive strength. For that, it is needed to improve the creep coefficient taking into consideration the various environmental conditions and for different concrete mix composition. In order to optimize and improve a model, many optimization models were developed in the literature [13–15]; however, Bayesian inference is used in this paper as being an appropriate tool for revising and updating design codes.

The objective of this study is to evaluate the long-term Eurocode 2 creep coefficient model and to improve it by implementing a correction coefficient to the model using Bayesian inference for various environmental conditions and concrete mix composition.

2. Database and Method

2.1. *Experimental database*

The database used in this study is the Northwestern University (NU) database which was established in 2010-2013. The tests in this database are performed under various environmental conditions and by using different concrete mix composition such as concrete compressive strength (f_{cm} relative humidity (RH), age at loading (t_0), sustained stress over the compressive strength at loading age $\sigma/f_{cm(t0)}$, etc.

2.2. *Eurocode 2 model*

The compliance is the total load-induced strain at age t per unit caused by a unit uniaxial sustained load applied since loading age t_0, and according to Eurocode 2 model (EC2) [10], it is given by equation (1).

$$J(t,t_0) = 1/E_{cmt0} + \varphi_{28}(t,t_0)/E_{cm28} \quad (1)$$

where E_{cmt0} is the modulus of elasticity of concrete at the time of loading t_0 (MPa), E_{cm28} is the mean modulus of elasticity at 28 days (MPa), and the dimensionless 28-days creep coefficient $\varphi_{28}(t,t_0)$ gives the ratio of the creep strain since the start of loading at the age t_0 to the elastic strain due to a constant stress applied at a concrete age of 28 days.

This study aims to update $\varphi_{28}(t,t_0)$ at long-term by inserting a correction coefficient, A, into the formula as shown in equation (2).

$$\varphi_{upd}(t,t_0) = A \times \varphi_{28}(t,t_0) \quad (2)$$

2.3. *Bayesian Inference*

Bayesian inference is a method of statistical inference in which Bayes' theorem is used to deduce properties about a population or probability distribution from data. Bayes' theorem in model form is written as shown in equation (3).

$$P(\Theta|data) = [P(data|\Theta)xP(\Theta)]/P(data) \quad (3)$$

where $\Theta = \{\mu, \sigma\}$ represents the set of parameters of a Gaussian distribution where μ is the mean and σ is the standard deviation. $P(\Theta)$ is the prior distribution and it represents the beliefs/knowledge about the true value of the parameters, $P(\Theta|data)$ is known as the posterior distribution. $P(data|\Theta)$ is the likelihood distribution. Therefore, the posterior distribution is calculated using the prior beliefs updated with the likelihood.

2.3.1. Prior for μ and $\sigma 2$. The Gaussian distribution is conjugate to itself with respect to a Gaussian likelihood function. Based on this theory, the conditional prior distribution for μ is considered as a normal with a prior mean of m_0 and a prior variance of σ^2/n_0 ($\mu|\sigma^2 \sim N(m_0,\sigma^2/n_0)$), and it is expressed as:

$$\boldsymbol{\mu|\sigma^2 \propto (n_0^{1/2}/\sigma\sqrt{2\pi}) \times exp\{-0.5n_0[(\mu - m_0)/\sigma]^2\}} \quad (4)$$

As σ^2 is unknown, a prior distribution should be used to describe the uncertainty about the variance before seeing the data. The variance is known to be positive, continuous, and with no upper limit, so a gamma distribution can be used as a prior distribution for variance. However, this choice does not lead to a posterior distribution in the same family or it is recognizable as any common distribution. To overcome this issue, the precision ϕ, defined as the inverse of the variance, is used as it has a conjugate gamma prior distribution. The conjugate prior for ϕ is expressed as:

$$\emptyset \sim Gamma(\vartheta_0/2, \vartheta_0 s_0^2/2) \quad (5)$$

where $\vartheta_0/2$ is the shape parameter, and $\vartheta_0 s_0^2/2$ is the rate parameter. In fact, the hyper-parameter ϑ_0 is interpreted as the prior degrees of freedom and the hyper-parameter s_0^2 as the prior variance or initial prior estimate for σ^2. Therefore, the conjugate prior for ϕ, $p(\phi)$ can be written as:

$$p(\emptyset) \propto \emptyset^{(\vartheta_0/2)-1} exp\{-(\vartheta_0 s_0^2 \emptyset)/2\} \quad (6)$$

The joint distribution for the pair (μ, ϕ) is then a normal-gamma distribution, NormalGamma(m_0, n_0, s_0^2, ϑ_0), with the four hyper-parameters m_0, n_0, s_0^2, and ϑ_0.

2.3.2. Likelihood distribution $\mathcal{L}(\mu, \sigma^2)$. Under the assumption that the data are a random sample of size n from a normal population with mean μ and variance σ^2, the sampling distribution of the data is the product of independent normal distributions with mean μ and variance σ^2.

$$p(A_1, \dots, A_n|\mu, \sigma^2) = \prod_{i=1}^{n} \frac{1}{\sqrt{2\pi\sigma^2}} e^{\left\{-\frac{1}{2}\left(\frac{(A_i-\mu)}{\sigma}\right)^2\right\}} \quad (7)$$

$$\mathcal{L}(\mu, \sigma^2) \propto \frac{1}{(2\pi\sigma^2)^{n/2}} exp\left\{-\left(\frac{\sum_{i=1}^{n}(A_i-\mu)^2}{2\sigma^2}\right)\right\} \quad (8)$$

After examining some calculations and considering $\bar{A}=(\sum_{i=1}^{n} A_i)/n$, the sample variance $s^2=(\sum_{i=1}^{n}(A_i-\bar{A})^2)/(n-1)$ and substituting σ^2 by 1/ϕ, the following equation is obtained:

$$\mathcal{L}(\mu, \emptyset) \propto \emptyset^{((n-1)/2)} \times exp\{-0.5\emptyset(n-1)s^2\} \times (n\emptyset)^{1/2} exp\{-0.5\emptyset n(\bar{A}-\mu)^2\} \quad (9)$$

2.3.3. Posterior distribution. The joint posterior distribution according to Bayes Theorem is proportional to the likelihood of the parameters times the joint prior distribution.

$$\boldsymbol{p(\mu, \phi|data) \propto L(\mu, \phi)p(\mu|\phi)p(\phi)}$$

(10)

After examining some calculations, and by considering $n_n = n + n_0$, $m_n = (n\bar{A} + n_0 m_0)/(n + n_0)$, $\vartheta_n = n + \vartheta_0$ and $s_n^2 = \left\{(n-1)s^2 + \vartheta_0 s_0^2 + \frac{nn_0}{n_n}[(m_0 - \bar{A})^2]\right\}/\vartheta_n$, the following equation is obtained:

$$p(\mu, \emptyset|data) \propto \emptyset^{0.5\vartheta_n - 1} exp\{-0.5\vartheta_n s_n^2 \emptyset\} \times (n_n \emptyset)^{0.5} exp\{-0.5\emptyset n_n(\mu - m_n)^2\} \quad (11)$$

Finally,

$$\emptyset|data \sim Gamma(0.5\vartheta_n, 0.5\vartheta_n s_n^2) \quad (12)$$

$$\mu|\emptyset, data \sim Normal(m_n, (\emptyset n_n)^{-1}) \quad (13)$$

3. Results and discussion

1488 observations, extracted from 179 experiments performed under various conditions with readings after 700 days of load application, are chosen from the NU database. For each observation, the creep coefficient is calculated according to EC2 model [10]. Finally, the ratio between observed and predicted values, A, are obtained for each observation. After exploring the data and to obtain more accurate results, the data is divided into six groups as shown in Table 1.

Table 1. Characteristics of each group

Group	Conditions
1	$f_{cm} \leq 35$ MPa and $\sigma/f_{cm(t_0)} < 0.3$
2	$f_{cm} \leq 33$ MPa and $\sigma/f_{cm(t_0)} \geq 0.3$
3	33 MPa < $f_{cm} \leq 35$ MPa and $\sigma/f_{cm(t_0)} \geq 0.3$
4	35 MPa < f_{cm} < 45 MPa and RH < 80%
5	$f_{cm} \geq 45$ MPa and RH < 80%

6	$f_{cm} \geq 45$ MPa and RH $\geq$ 80%

3.1. Bayesian Inference
Readings from the same experiment at different times and under the same condition cannot be independent of each other. However, it may be reasonable to assume that the ratio between the creep coefficients observed and predicted according to EC2 in random samples of concrete under various conditions and at different times are independent of each other. The normality of the data in the six groups are verified using the Shapiro-Wilk test.

A collection of random variables (A_1, A_2 ... A_n) is assumed to be independent and identically distributed from a normal population with mean μ and variance σ^2. Bayes Theorem leads to the posterior distribution for μ and σ2 given the observed data and expressed as:

$$p(\mu, \sigma^2 | A_1, \ldots, A_n) = \frac{p(A_1, \ldots, A_n | \mu, \sigma^2) \times p(\mu, \sigma^2)}{normalizing\ constant} = \frac{\mathcal{L}(\mu, \sigma^2) \times p(\mu, \sigma^2)}{normalizing\ constant} \tag{14}$$

3.1.1. Likelihood distribution $\mathcal{L}(\mu, \sigma^2)$. The variable A in six groups has high skewed and non-normal distribution and the histogram shows an exponential distribution. Therefore, a natural log transformation of A is applied where each variable of A is replaced by y = ln(A). The normality of the data in the six groups is verified using Shapiro-Wilk's test.

3.1.2. Prior for μ *and* σ^2. The correction coefficient is always positive. Therefore, A is lognormally distributed. In this case, the natural logarithm of A, ln(A), is normally distributed. The prior hyper-parameters used are assumed as follows: $m_0 = 1$, $s_0^2 = 0.12$, $n_0 = 60$, and $v_0 = 59$.

3.1.3. Posterior distribution. As shown in equation (13), the posterior mean, m_n, is calculated and the values of the correction coefficient for the six groups are shown in Table 2. Therefore, EC2 model underestimates the creep coefficient for the groups 2, 4 and 5, and overestimates it for the groups 1, 3 and 6.

Table 2. Values of the correction coefficient A.

	Group 1	Group 2	Group 3	Group 4	Group 5	Group 6
A	0.83	1.39	0.8	1.21	1.3	0.94

In this study, the statistical indicators used are BP coefficient of variation ($\overline{w}_{BP}$ %) [16], CEB coefficient of variation (V_{CEB}) [17], and the Gardner coefficient of variation (ω_G) [18]. These indicators are calculated before and after the implementation of the correction coefficient in EC2 creep coefficient model (Table 3). As shown in Table 3, the implementation of the correction coefficient A provides a significant improvement on the results where $\overline{w}_{BP}$, V_{CEB}, and ω_G have decreased by 22%, 22%, and 25%, respectively.

Table 3. Statistical indicators of the Eurocode 2 creep coefficient model before and after the implementation of the correction coefficient A

	$\overline{w}_{BP}$	V_{CEB}	ω_G
Before implementation of A	0.58	0.50	0.57
After implementation of A	0.45	0.38	0.43

4. Conclusion

1488 observations, with readings after 700 days of load application, were selected from the NU database to improve the EC2 creep coefficient prediction by implementing a correction coefficient into the model. After exploring the data, these observations were divided into six groups, and the correction coefficients for the six groups were calculated using Bayesian inference.

The results shows that the EC2 model was significantly improved where the statistical indicators $\overline{w}_{BP}$, V_{CEB}, and ω_G have been decreased by 22%, 22%, and 25%, respectively.

References

[1] Daou H, Raphael W, Chateauneuf A and Geara F 2019 Probabilistic Assessment of Structural Safety of Complex Structures - Application to Terminal 2E at Roissy, CDG Airport *Procedia Structural Integity.* **22** 17–24
[2] Daou H and Raphael W 2020 Investigation of the Significance of Factors Affecting Long-Term Creep of Concrete using Screening Designs *IOP Conference Series: Materials Science and Engineering* **809**
[3] Daou H and Rapahel W 2021 Identifying the Weight of Factors Affecting Creep of Concrete Using Factorial ANOVA *Key Engineering Materials* **872** 21–5
[4] Saad L, Chateauneuf A and Raphael W 2018 Robust formulation for Reliability-based design optimization of structures *Structural and Multidisciplinary Optimization* **57** 2233-2248
[5] Daou H and Raphael W 2021 Comparison between Various Creep Calculation Methods for the Time-dependent Analysis of Terminal 2E at Roissy *Jordanian Journal of Civil Engineering* **15** 64–76
[6] Raphael W, Zgheib E and Chateauneuf A 2018 Experimental investigations and sensitivity analysis to explain the large creep of concrete deformations in the bridge of Cheviré *Case Studies in Construction Materials* **9** e00176
[7] Daou H, Abou Salha W, Raphael W and Chateauneuf A 2019 Explanation of the collapse of Terminal 2E at Roissy–CDG Airport by nonlinear deterministic and reliability analyses *Case Studies in Construction Materials* **10**
[8] Raphael W, Faddoul R, Feghaly R and Chateauneuf A 2012 Analysis of Roissy Airport Terminal 2E collapse using deterministic and reliability assessments *Engineering Failure Analysis* **20** 1–8
[9] Chateauneuf A, Raphael W, Rostand J B and Pitti M 2014 Reliability of prestressed concrete structures considering creep models *Structure and Infrastructure Engineering: Maintenance, Management, Life-Cycle Design and Performance* **10:12** 1595–605
[10] CEN 2004 EN 1992-1-1, Eurocode 2: Design of concrete structures - Part 1-1 : General rules and rules for buildings
[11] Raphael W, Faddoul R, Geara F and Chateauneuf A 2012 Improvements to the Eurocode 2 shrinkage model for concrete using a large experimental database *Structural Concrete* **13** 174–81
[12] Raphael W, Faddoul R, Selouan D E-A and Chateauneuf A 2009 Information-based formulation for Bayesian updating of the Eurocode 2 creep model *Structural Concrete* **10** 1–55
[13] Faddoul R, Soubra A, Raphael W and Chateauneuf A 2013 Extension of dynamic programming models for management optimisation from single structure to multi-structures level *Structure and Infrastructure Engineering: Maintenance, Management, Life-Cycle Design and Performance.* **9:5** 432–47
[14] Faddoul R, Raphael W, Soubra A-H and Chateauneuf A 2013 Incorporating Bayesian Networks in Markov Decision Processes *Journal of Infrastructure Systems* **19** 415–24
[15] Faddoul R, Raphael W and Chateauneuf A 2018 Maintenance optimization of series systems subject to reliability constraints *Reliability Engineering and Systen Safety* **180** 179–88
[16] Bažant Z P and Panula L 1978 Practical prediction of time-dependent deformations of concrete. Part II: Basic creep *Matériaux et Constructions* **11** 317–28
[17] Müller H S and Hilsdorf H K *Evaluation of the time-dependent behaviour of concrete: summary report on the work of the general task force group No. 199* (Lausanne, Switzerland)
[18] Gardner N J 2004 Comparison of prediction provisions for drying shrinkage and creep of normal-strength concretes *Canadian Journal of Civil Engineering* **31** 767–75

Unconfined Compressive Strength of Rehabilitated Road Base via Polymer-Modified Cold In-Place Recycling (CIPR)

Poi-CheongTan[1*], Chee-Ming Chan[2]

[1, 2]Faculty of Engineering Technology, Universiti Tun Hussein Onn, Malaysia.

E-mail: hn180028@siswa.uthm.edu.my

Abstract. Cold In-Place Recycling (CIPR) is a road maintenance technology, which chemically stabilize, recycle and improve existing asphalt concrete together with road base underneath. Cement is the most common chemical for CIPR adopted in Malaysia. However, shrinkage crack remains as unresolved technical drawback of cement stabilization, road practitioners are always seeking alternative materials to replace cement in CIPR. A proprietary product, Polymer Modified Cementitious (PMC) chemical binder was identified as the research object to verify the suitability in CIPR application and to study the strength-gain over time, through laboratory testing. To replicate CIPR process, Crusher run (CR) and Recycled Asphalt Pavement (RAP) were mixed at various ratio to study the effect of CR to RAP ratio on strength gained; with PMC added in dosage between 2-4% per dry weight of the materials to compare strength gain at different dosage. Samples were compacted and cured for 7 days then subjected to Unconfined Compressive Test where the strength UCS (q_{u7}) was recorded. Strength-gain over time from 1 day to 28 days, for a specific CR-RAP ratio at fixed PMC dosage was also studied. It was found that UCS was directly proportionate to PMC dosage and amount of CR component in the mixture. This study concluded that at various combination of CR-RAP ratio, with varying PMC dosage, minimum requirement of UCS (7-day) of 2 MPa for road base application specified by local authority is achievable. The correlation between strength and curing duration obtained serves as guideline to extrapolate long-term strength from short-term strength.

1. Introduction

Soil stabilization is a common engineering method to treat unsuitable materials or recycle existing materials, and turn into usable materials, with desirable properties. In general, soil stabilization can be categorized into mechanical stabilization and chemical stabilization [1]. Means of mechanical stabilization includes compaction (densification), drainage [2], altering particle size distribution, replacement, consolidation/dewatering and reinforcement [3]. Chemical stabilization is the method to stabilize soil through chemical reaction. In general, soil properties such as strength, stiffness soil permeability, compressibility, plasticity and swelling/shrinkage potential of soil can be improved through chemical stabilization.

History of applying chemical stabilization for road construction in Malaysia was as early as 1980s'. In Malaysia civil construction industry context, in-situ chemical stabilization process involving stabilization of existing asphalt concrete together with underlying materials with addition of stabilizing agent, is often known as Cold In-Place Recycling (CIPR). Cement is a most common chemical used in CIPR in Malaysia. However, with shrinkage crack inherited as natural characteristic of cement stabilized bases, overall performance of cement stabilized base is remain a concern [4]. Therefore, there is a need to evaluate other potential chemical binder as alternative.

This study aims to evaluate a proprietary polymer modified cementitious (PMC) chemical, and verify as the alternative chemical binder in CIPR process in terms of mechanical strength. Through laboratory experiments optimum chemical dosage is verified, and thus serves a guideline for field application. Besides, Unconfined compressive strength (UCS) with various curing during is also studied to understand strength development pattern over time.

2. Design of Experiment

Unconfined compressive strength is a direct indication of a material's strength to resist deformation under unconfined condition. It is one of the common design criteria of chemical stabilization adopted by Malaysia Public Works Department (JKR Malaysia). Therefore in this study, unconfined compressive strength is only criteria to evaluate the technical feasibility of the PMC in road base stabilization. UCS tests are carried out using cylindrical sample in accordance to BS 1924:1990 (Part 2).

This study is carried out in two stages, namely Stage I-Verification Technical Suitability of PMC, and Stage II-Strength-Gain Over Time. Stage I experiment aimed to determine the PMC dosage and crusher run-RAP mixing ratio, which attain minimum strength required. During stage I, curing duration was kept constant at 7-day. Duration of 7-day was chosen as the reference because this is the standard curing duration specified in requirement of Unconfined Compressive Strength Test (UCS) in Standard Specification for Road Works Section 4: Flexible Pavement and Section 18: Soil Stabilization published by JKR Malaysia [5]. Nevertheless, as opposed to cylindrical samples used in this study, cube UCS samples were specified as test specimen in both specification. Therefore, UCS of both cylindrical and cube samples prepared at a specific condition, were first tested to verify correlation of UCS results between both types of sample.

Table 1 summarizes ratio of various CR to RAP mixture, ranging from 100% CR to 100% RAP, to study the effect of mixing ratio. PMC dosage was set at 2%, 3% and 4%, to investigate correlation of strength development against dosage.

Table 1. Ratio of crusher run and RAP mixture.

Sample Reference	Crusher Run Percentage (%)	RAP Percentage (%)
C100R0	100	0
C75R25	75	25
C50R50	50	50
C27R75	25	75
C0R100	0	100

Stage II experimental work aimed to study time-dependency of strength development with the PMC chemical stabilization. Adopting the optimum PMC dosage mixing and specific ratio of crusher run-RAP and obtained in Stage I where these two factors were set as constants, the curing duration was varied between 1-day, 3-day, 14-day and 28-day. From the strength development pattern obtained in Stage II, the short-term strength could be extrapolated to predict the long-term strength.

Since this study is limited to laboratory experiment, recycle asphalt pavement (RAP, as knows as milling waste) and crusher run were used to replicate asphalt concrete layer and crusher road base layer in road rehabilitation. The PMC used is a commercialised product, supplied by industrial collaborator from Malaysia. The product has been applied in Malaysia road construction for more than five years. This proprietary product consists of cementitious chemical, ionic chemical and polymeric fiber.

3. Results and Discussions

In order to relate UCS obtained in this study, which was using cylindrical sample, to the requirement specified by JKR Malaysia, correlation between UCS cylindrical (q_{u7}) sample and UCS cube ($q_{u7,cube}$) sample, is to be obtained. Refer to BS EN1992-1-1:2004, the strength of cylindrical concrete approximates 80% of cube concrete. Unconfined compressive strength test was carried out on 2 sets C100R0 with 3% PMC dosage samples, in cylindrical (100 mm diameter, 200 mm height) and cube samples (100 mm x 100 mm x100 mm) respectively to verify the correlation. The samples were cured for seven days prior to the tests. Table 2 summaries results of the UCS tests of both sets samples. The ratio of q_{u7} /$q_{u7,cube}$ was found to be 0.78, close to the ratio found in BS EN1992-1-1:2004. The minimum requirement of UCS (7-day) stipulated by JKR Malaysia in the aforementioned specification was 2 MPa using cube sample. Therefore one can estimate the minimum UCS (7-day) required is 80% of 2 MPa, which equals to 1.6 MPa.

Table 2. Comparison of UCS Using Cylindrical and Cube Sample

Sample No	q_u (MPa)	$q_{u,cube}$ (MPa)	Ratio q_u /$q_{u,cube}$
S1	--[a]	3.22	
S2	2.38	3.22	
S3	2.58	3.10	
Average	2.48	3.18	0.78

[a] due to experimental error, f_{ck} of S1 was not considered in the calculation for average q_u

3.1. *Stage I – Verification Technical Suitability of PMC*

Figure 1 shows the relationship between UCS and PMC dosage for various CR-RAP ratio. It was found that UCS attained is direct proportionate to PMC added in linear model. With 2% of PMC dosage, UCS (7-day) of 1.6 MPa was not achievable for all CR-RAP ratio. Therefore, dosage of 2% or below is not advisable for road base rehabilitation. Nevertheless, UCS of C100R0 at 2% dosage obtained were 1.50 MPa, 1.39MPa and 1.51 MPa, which were close to 1.60 MPa, more study is required to further verify suitability to apply 2% dosage for C100R0.

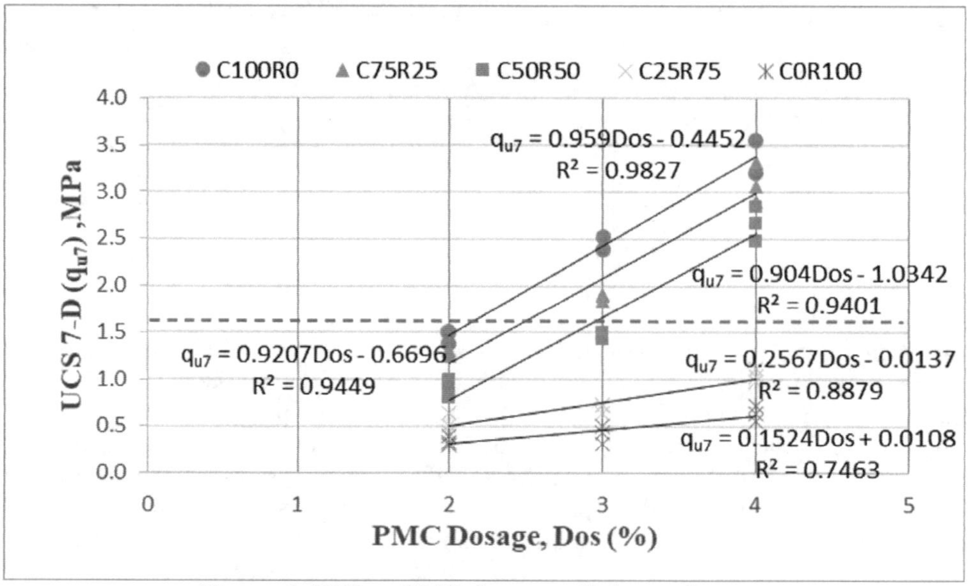

Figure 1. UCS vs PMC dosage at 7-day curing.

By comparing trendlines obtained in Figure 1, trendline for C100R0 series has highest magnitude of gradient, which indicated effect of increasing of dosage on increment of UCS is higher for C100R0. A decresing trend of the trendline gradient is observed as ratio of CR is decreasing. The gradient decreased greatly after C50R50 series, indicated PMC is less effective to improve UCS when RAP ratio is more than 50%. In order to further study relationship of UCS and CR-RAP ratio, the data was rearranged and presented in Figure 2. It is apparent that with PMC stabilization, resultant UCS is directly proportionate to percentage of CR component in the mixture in linear model. By comparing the two factors, PMC dosage and CR ratio, which affect UCS, CR ratio is more dominant factor. Average UCS of C100R0 at 2% PMC dosage was 1.47 MPa, while C0R100 even at higher dosage of 4%, average UCS of 0.64 MPa was recorded.

Besides, with increasing of PMC dosage, the effect on increasing of UCS, in terms of both absolute value and percentage, is more prominent at higher CR ratio. For C100R0 samples, by increasing PMC dosage from 2% to 4%, average UCS increased from 1.47 MPa to 3.38 MPa, an increment of 1.91 MPa or equivalent to 131%. With same increment of PMC dosage from 2% to 4%, average UCS of C0R100 increased from 0.34 MPa to 0.64 MPa, an increment of 0.30 MPa or equivalent to 90%. From technical point of view, it is advisable to maximize the CR ratio in order to maximize the resultant strength achievable. On the other hand, from environmental and sustainability point of view, it is more beneficial to recycle more RAP in the mixture. Therefore one has to strike the balance to decide CR-RAP ratio with the PMC dosage, in order to optimize resultant UCS.

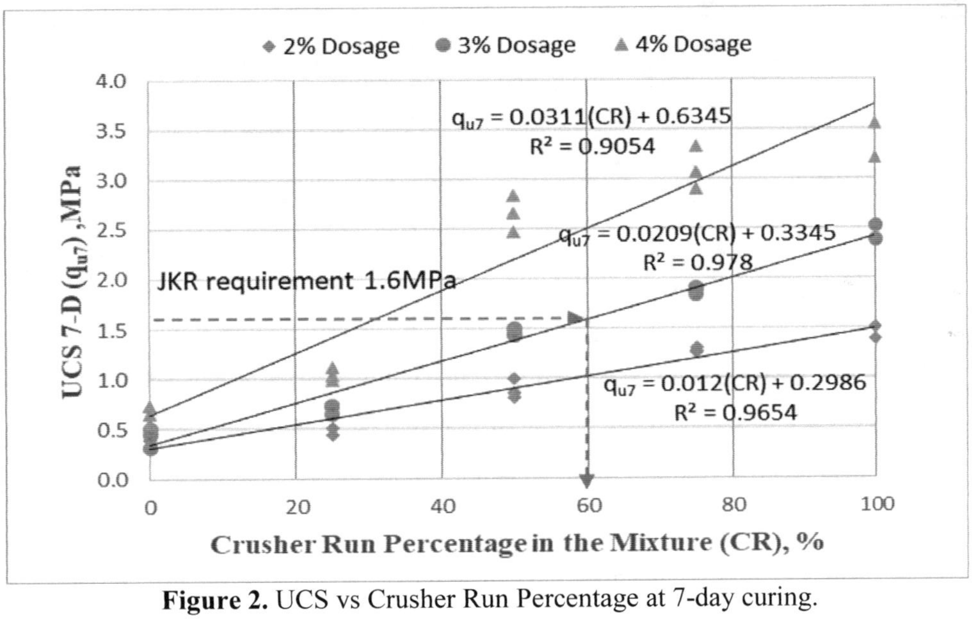

Figure 2. UCS vs Crusher Run Percentage at 7-day curing.

Figure 2 showed that UCS requirement of 1.6 MPa is achievable with various CR-RAP ratio, applying different PMC dosage. Depending on thickness of existing asphalt concrete layer to be rehabilitate, the ratio of CR to RAP is known. Then the PMC dosage required can be estimated from Figure 2. In Stage II time-dependent of UCS development, the ratio 60% CR to 40% RAP (C60R40) was selected as the experiment standard, in order to replicate the situation where in-situ chemical stabilization involves mixing of 120mm of existing asphalt concrete layer and 180mm existing crusher run road base. With the correlation established, the PMC dosage of 3% is required to meet UCS (7-day) requirement of 1.6 MPa.

It was suggested that participle size distribution of CR and RAP would affect resultant UCS. Sieve analysis was carried out for both pure CR and pure RAP respectively verify this hypothesis, by studying difference in particle size distribution between both samples. Sieve analysis was presented against sieve envelop on road base materials approved by JKR Malaysia, plotted as the lower and upper limits in Figure 3.

RAP is the by-product of asphalt concrete milling process, which the cutting drum rotates the cutting tools in high speed across existing asphalt concrete surface and breaks the asphalt concrete with mechanical force. Therefore grain size distribution of RAP is uncontrolled. Refers to Figure 3, for C0R100, particles passed 0.118mm sieve and below did not meet the lower limit of the acceptable sieve envelop. The finding implied C0R100 lacked of fines which filled up the voids in between particles during compaction, leading to poor bonding, less mechanical interlocking between particles, and lower compactability. During compressive test, more voids within the sample act as failure surface, resulting in lower resultant UCS. Unlike RAP, CR is purposely crushed and mixed with better grain size distribution, to achieve certain requirement of sieve envelop. The more CR percentage in the mixture, the better overall grain size distribution is, and thus improve resultant UCS after PMC stabilization.

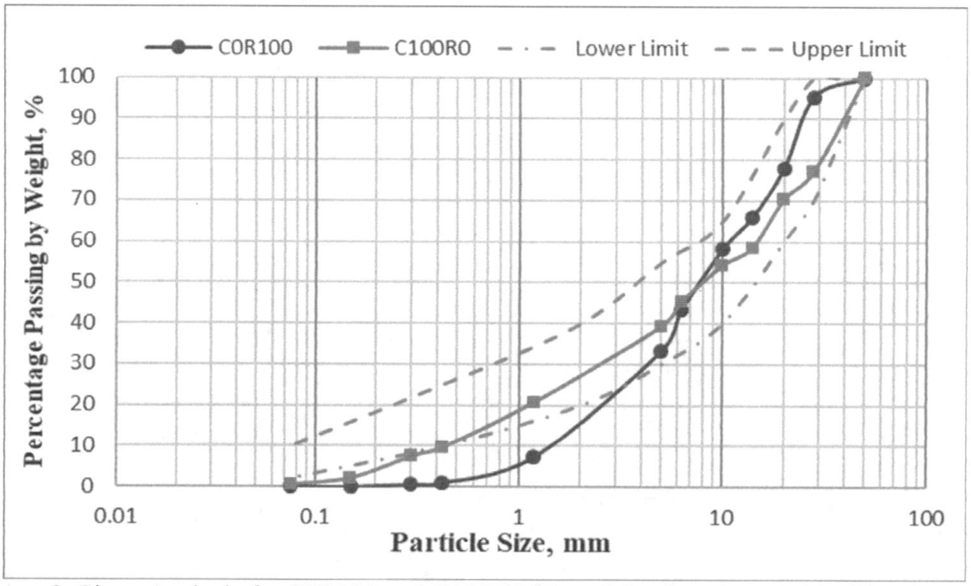

Figure 3. Sieve Analysis for C0R100 and C100R0 Plotted Against Acceptable Sieve Envelop

To study effect of grain size distribution on UCS, a set of C0R100 samples, treated with PMC at dosage of 3%, was prepared, by artificially remix the particles from difference sieve size, in accordance to grain size distribution of C100R0. The samples were compacted and cured for 7 days, then subjected to unconfined compressive test. Table 3 summarizes and compares UCS results of remix C0R100 against C0R100 of original grain size distribution, as well as C100R0, with same PMC dosage and curing duration. It was found that, by changing the grain size distribution of RAP, resultant UCS results were improved. However, even with identical grain size distribution with CR, UCS attained by RAP were still lower CR. This finding implies there is/are intrinsic difference(s) between RAP and CR which affect(s) resultant strength after PMC treatment.

Table 3. UCS (7-day) of C100R0 and C0R100 Treated with 3% PMC

Sample No	C100R0 (MPa)	C0R100 (original) (MPa)	C0R100 (remix) (MPa)
S1	1.86	0.30	0.64
S2	2.38	0.39	0.47
S3	2.52	0.33	0.65

3.2. *Strength-Gain Over Time*

UCS attained against curing duration was presented in Figure 4. Strength of PMC stabilization develops over time in logarithmic model, which is similar to strength-gain curve of a concrete. With the strength development equation obtained, the relationship of UCS at various curing duration against UCS at 7-day curing and 28-day curing is summarized in Table 4. Curing duration of 7-day is always used as design requirement. With the relationship established, UCS can be tested at early stage to estimate the UCS achievable at 7-day. Thus, any non-conforming work can be identified and thus rectified at earlier stage.

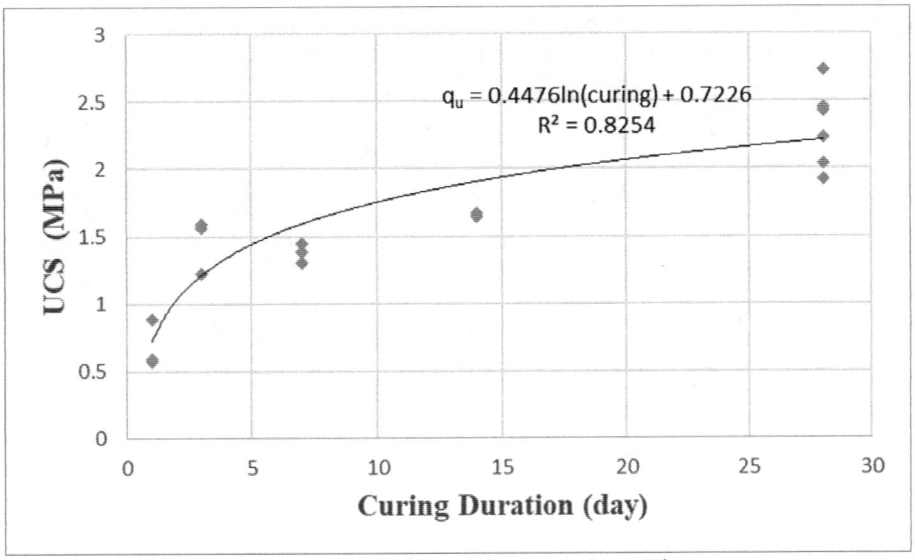

Figure 4. Strength Development Over Time

Table 4. Ratio of UCS at Various Curing Duration Against UCS 7-day and UCS 28-day

Curing Duration	Ratio $q_{u\ x\text{-}day} / q_{u28}$	Ratio $q_{u\ x\text{-}day} / q_{u7}$
1-day	33%	45%
3-day	55%	76%
7-day	72%	100%
14-day	86%	119%
28-day	100%	140%

4. Conclusions

Unconfined compressive strength of PMC stabilized CR-RAP mixture is a direct proportionate to PMC dosage added, as well as the ratio of crusher run component in the mixture. Ratio of crusher run in the mixture is the more dominant factor deciding resultant UCS. It is verified that, min UCS (7-day, cube sample) requirement of 2 MPa stipulated by JKR Malaysia for road base chemical stabilization usage is achievable with various combination of CR-RAP ratio, and PMC dosage. With 40% substitution of CR by RAP (C60R40), PMC dosage of 3% is adequate to comply to JKR Malaysia requirement. From this study, strength-gain over time for C60R40 is established, to serve as guideline to extrapolate and estimate long term strength. With correlation obtained in this study, one has to optimize the CR-RAP ratio (existing pavement design an condition) and PMC dosage (commercial consideration), which is technically, commercially and environmentally viable.

References

[1] Patel A and Patel A 2019 Soil Stabilization *Geotechnical Investigations and Improvement of Ground Conditions* 19-27
[2] Kitazume M and Terashi M 2013 *The Deep Mixing Method* (London: CPC Press) 6
[3] Makusa G P 2013 *State of the Art Review Soil Stabilization Methods and Materials in Engineering Practice* (Luleå: Luleå University of Technology) 1
[4] Louw S, Jones D and Hammack J 2016 *Pavement Recycling: shrinkage crack mitigation in cement-treated pavement layers - Phase 1 Laboratory Testing* (Davis: Institute of Transportation Studies, University of California)
[5] Malaysian Public Works Department 2019 Section 18: Soil Stabilization. In *Standard Specification for Road Works* S18-8, Firoozi A, Guney Olgun C, Firoozi A, Baghini M S 2017 Fundamentals of soil stabilization *International Journal of Geo-Engineering* **8**

Development of a model for predicting dynamic response of a sphere at viscoelastic interface: A dynamic Hertz model

H Koruk[1]

[1]Mechanical Engineering Department, MEF University, 34396 Istanbul, Turkey

E-mail: korukh@mef.edu.tr

Abstract. A model for predicting the dynamic response of a sphere at viscoelastic interface is presented. The model is based on Hertz contact model and the model for a sphere in a medium. In addition to the elastic properties of medium and the size of sphere, the model considers the density of sphere, the density and viscosity of medium, and damping of oscillations of sphere due to radiation of shear waves. The model can predict not only the effects of the mechanical properties of medium, the physical properties of sphere, and the amplitude of excitation force on sphere displacement, but also the effects of these parameters on shift of resonance frequency. The proposed model can be used to identify the elastic and damping properties of materials, and to understand the dynamic responses of spherical objects at viscoelastic interfaces in practical applications.

1. Introduction

Spherical objects, such as bubbles and spheres, embedded in mediums and at viscoelastic interfaces are encountered in many applications [1-3]. Spherical objects embedded in a material have been used to determine material properties, such as the elasticity modulus [1,4]. However, in practice, for example, in therapeutic and diagnostic ultrasound applications that use microbubbles [5] and in atomic force microscopy or indentation tests that use spheres [6-8], the spherical objects are at viscoelastic interfaces.

More recently, the use of microbubbles in a fluid to push against tissue under ultrasound exposure was proposed to improve the contrast and spatial resolution of elasticity imaging [9]. In addition, mathematical models for the displacements of a bubble located at medium interfaces in response to external loads were proposed [10,11] and these models were evaluated experimentally [12]. The Hertz model, or modified Hertz models, are widely used in various applications, including atomic force microscopy and nanoindentation, to predict the responses of spheres at medium interfaces under external loads [6,8,13]. However, the models in the literature for material identification are based on some fictive arrangements of springs, do not take into account the effect of the sphere mass for dynamic loading and radiation damping, or require the use of a variety of parameters including relaxation times under constant load and deformation [14,15]. Different models have been proposed to investigate the nonlinear vibrations of Hertzian contact, however, they are based on a mass-spring-damper system [3,16]. There are some models for studying dynamic-contact stiffness at the interface between a vibrating rigid sphere and a semi-infinite viscoelastic solid [17]. Overall, most studies in the literature focus on the steady-state deformation of the sphere or relaxation [18,19]. There are some studies on the dynamic response analysis of a spherical subject impacting a metal plate [20]. It should be noted that different approaches are used to handle the stiffness of the system consisting a sphere in contact with an elastic body [21]. There are some studies in which the oscillatory response is investigated, however, the sphere is attached to an auxiliary system such as a pendulum in these studies [22,23]. A practical model to analyse the dynamic response of a sphere at viscoelastic interface, considering wave propagation using a continuum linear-viscoelastic medium and the Hertz contact model, will be invaluable, especially for material identification purposes.

In this paper, a model for the dynamic response of a sphere at viscoelastic interface exposed to an external dynamic load is presented. The model is based on the Hertz contact model [13] and the model for the sphere embedded in a medium [24-25]. In addition to the elastic properties of medium and the size of sphere, the model takes into account the density and viscosity of medium and the mass of sphere. In the model, the damping of the oscillations of the sphere due to the radiation of shear waves is taken into account. The proposed model was assessed using the model of the sphere embedded in a medium. The results show that the model can predict not only the effects of the mechanical properties of medium, the physical properties of sphere and the amplitude of the excitation force on the displacement of the sphere, but also the effects of these parameters on the shift of the resonance frequency. The model proposed in this study can be used to identify the elastic and damping properties of materials, and to understand the dynamic responses of spherical objects at viscoelastic interfaces in practical applications

2. Mathematical model

The spheres embedded in a viscoelastic medium and at a viscoelastic interface are shown in Figure 1. The model for the dynamic response of the sphere at the viscoelastic interface here was developed based on the Hertz contact model [13] and the model for a sphere embedded in a medium [24,25]. It is assumed that the sphere is always in contact with the medium.

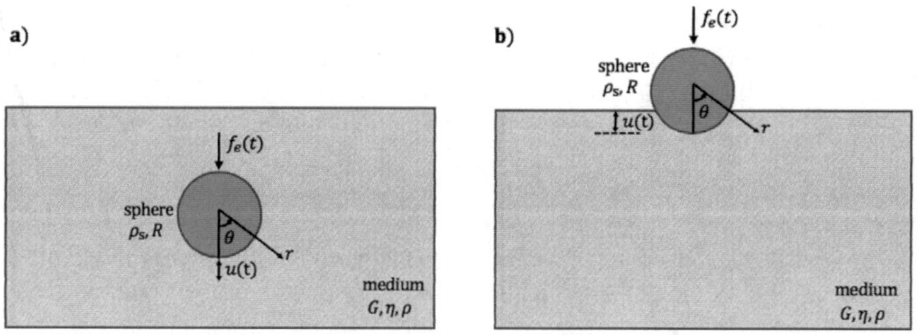

Figure 1. The spheres embedded in a viscoelastic medium (a) and at a viscoelastic interface (b).

The frequency-domain equation that couples the displacement of the sphere embedded in an elastic medium for a dynamic external force is as follows [24]:

$$F_e = -m_s\omega^2 U + 6\pi GRU\left(1 - jkR - \frac{1}{9}k^2R^2\right) \tag{1}$$

where m_s and R are the mass and radius of sphere, G is the shear modulus of medium, $k = \frac{\omega}{\sqrt{G/\rho}}$ is the wave number of the shear wave with the frequency ω, ρ is the density of medium, $j = \sqrt{-1}$ and F_e and U are the Fourier transforms of the external force $f_e(t)$ and displacement $u(t)$, respectively. The mass of a sphere is given by $m_s = \frac{4}{3}\pi R^3 \rho_s$ where ρ_s is the density of the sphere. Aglyamov et al. [25]

considered the external force as a rectangular pulse with the amplitude of f_0 and a duration of τ (i.e., the constant force f_0 is applied for a short time τ and then removed). It should be noted that the response to the rectangular pulse simulates the impulse response for small τ values and the step response for large τ values with $0 \leq t \leq \tau$. It can be calculated that the Fourier transform of the rectangular pulse is $F_e = -\frac{jf_0}{\omega}\left(e^{j\omega\tau} - 1\right)$. Aglyamov et al. [25] replaced the shear modulus G with $(G - j\omega\eta)$ in equation (1), and hence they obtained the equation that couples the displacement of the sphere in a viscoelastic medium for a dynamic external force as follows:

$$U = \frac{(-jf_0/\omega)(e^{j\omega\tau}-1)}{6\pi(G-j\omega\eta)R\left[1-jkR-\frac{1}{9}(1+2\beta)k^2R^2\right]} \tag{2}$$

where $\beta = \rho_s/\rho$. By using the inverse Fourier transform, $u(t) = \mathcal{F}^{-1}\{U\}$, the time-domain response of the sphere in the viscoelastic medium can be found as [25]:

$$u(t) = -\frac{jf_0}{12\pi^2R}\int_{-\infty}^{\infty}\frac{(e^{j\omega\tau}-1)e^{-j\omega t}}{\omega(G-j\omega\eta)\left(1-jkR-\frac{1}{9}(1+2\beta)k^2R^2\right)}\mathrm{d}\omega \tag{3}$$

Using $k^2 = \frac{\omega^2}{G/\rho}$, the last term in equation (1) becomes $-6\pi GRU\frac{1}{9}k^2R^2 = -\frac{2}{3}\pi\rho R^3\omega^2U$. Hence, equation (1) can be written as follows:

$$F_e = \frac{4}{3}\pi R^3\rho_s(-\omega^2U) + \frac{2}{3}\pi R^3\rho(-\omega^2U) + 6\pi GRU - 6\pi GRUj\omega\sqrt{\frac{\rho}{G}}R \tag{4}$$

The first term, $\frac{4}{3}\pi R^3\rho_s(-\omega^2U)$, in equation (4) or the corresponding term $\frac{4}{3}\pi R^3\rho_s\ddot{u}$ in the time domain shows the inertia force due to the mass of the sphere. It should be noted that the inertia force due to the mass of the sphere at the viscoelastic medium will be the same as the one for the sphere in a viscoelastic medium, i.e., $F_{\text{i,sphere}} = \frac{4}{3}\pi R^3\rho_s(-\omega^2U)$. The second term, $\frac{2}{3}\pi R^3\rho(-\omega^2U)$, in equation (4) or the corresponding term $\frac{2}{3}\pi R^3\rho\ddot{u}$ in the time domain is the inertia force due to the mass of the elastic medium involved in motion ($m_a = \frac{2}{3}\pi R^3\rho$ is the induced mass). It is noted that, depending on some other parameters including the submergence depth, the added mass can be much less than the mass of the sphere [26]. Therefore, based on the results in the literature [27,28], it is assumed that the inertia force due to the mass of the elastic medium involved in motion for the sphere at the medium interface is half of that of the sphere embedded in a medium, i.e., $F_{\text{i,medium}} = \frac{1}{3}\pi R^3\rho(-\omega^2U)$. Meanwhile, it can be shown that the effect of the induced mass is low and the mass of the sphere mainly dominates inertia forces, even for a sphere embedded in a medium. The m_s/m_a ratio is more than 15 for practical applications (i.e., for the steel sphere with $\rho_s = 7800$ kg/m^3 embedded in the gel with $\rho = 1000$ kg/m^3).

The third term, $6\pi GRU$, in equation (4) is the equivalent force related to the stiffness of the system. The fourth term in equation (4), being the product of the equivalent force related to the stiffness of the system and the term $-j\omega\sqrt{\frac{\rho}{G}}R$, denotes the damping of the oscillations of the sphere due to the radiation of shear waves. As can be seen, the damping due to shear wave propagation for the sphere embedded in a medium depends on the stiffness of the system, the displacement induced, the frequency (or wave number), and the sphere radius. The same parameters should also affect the damping term due to shear wave propagation for the sphere at a medium interface. Therefore, firstly, the correct stiffness term should be determined for the model of the sphere at a medium interface.

The Hertz theory describes the contact mechanics between the elastic sphere and an elastic half-space. In the Hertz model, the displacement (y) induced by the elastic sphere is correlated with the external force (f) with $f=\frac{4E^*\sqrt{R}}{3}y^{3/2}$ where E^* is the reduced Young's modulus, computed as $1/E^*=(1-\nu_{\text{sphere}}^2)/E_{\text{sphere}}+(1-\nu^2)/E$ and R is the relative radius [13]. It is known that the Hertz model accurately predicts the displacement of the sphere at an elastic half-space under static loading. Therefore, the displacement-force relationship for the sphere at a medium interface can be taken from Hertz theory. As opposed to the constant stiffness component for the sphere embedded in a medium, the stiffness component for the Hertz model is a function of applied force or induced displacement. Using the Hertz model, we can determine the stiffness as a function of sphere displacement as $k=\frac{df}{dy}=1.5\left(\frac{4E^*\sqrt{R}}{3}\right)y^{1/2}$. As it is well known, the stiffness changes from zero (at the beginning of the deformation) to the maximum value $k=1.5\left(\frac{4E^*\sqrt{R}}{3}\right)u^{1/2}$ for a displacement of u. The use of this stiffness and the aforementioned inertia force provides that the frequency of oscillation of the sphere at the viscoelastic interface is kept in accordance with Hertz theory which produces accurate estimations. In this study, such a model for the sphere at the viscoelastic interface is developed so that, not only the frequency of oscillations, but also the steady-state displacement of the sphere is the same as that of the Hertz-model; note that the accuracy of the Hertz model has already been validated. For this purpose, the effective stiffness of the sphere at an interface is defined as $k_1=k+\alpha$ where α is a constant stiffness component. Hence, for the force of amplitude f producing a steady-state displacement of u, we should have $f=k_1u=(k+\alpha)u=\left[1.5\left(\frac{4E^*\sqrt{R}}{3}\right)u^{1/2}+\alpha\right]u$. From this, we find $\alpha=-0.5\frac{f}{u}$, hence, we have $k_1=1.5\left(\frac{4E^*\sqrt{R}}{3}\right)u^{1/2}-0.5\frac{f}{u}$. As we can write $f^{1/3}=\left(\frac{4E^*\sqrt{R}}{3}\right)^{1/3}u^{1/2}$, it becomes $k_1=1.5\left(\frac{4E^*\sqrt{R}}{3}\right)^{2/3}f^{1/3}-0.5\frac{f}{u}$. Now, we can write the equation of motion for the sphere at an interface in time domain as:

$$f=\left(\frac{1}{3}\pi R^3\rho+\frac{4}{3}\pi R^3\rho_s\right)\ddot{u}+\left[1.5\left(\frac{4E^*\sqrt{R}}{3}\right)^{2/3}f^{1/3}-0.5\frac{f}{u}\right]u+f_r \tag{5}$$

where f_r is the force component related to the damping of the oscillations of the sphere due to the radiation of shear waves for the sphere at the viscoelastic medium interface. Equation (5) can be arranged as:

$$1.5f=\frac{1}{3}\pi R^3(4\rho_s+\rho)\ddot{u}+1.5\left(\frac{4E^*\sqrt{R}}{3}\right)^{2/3}f^{1/3}u+f_r \tag{6}$$

Equation (6) in frequency domain can be written as:

$$1.5(-jf_0/\omega)\left(e^{j\omega\tau}-1\right)=\frac{1}{3}\pi R^3(4\rho_s+\rho)(-\omega^2U)+1.5f_0^{1/3}\left(\frac{4E^*\sqrt{R}}{3}\right)^{2/3}U+F_r \tag{7}$$

It should be noted that, for the sphere at a viscoelastic interface, as the upper half-space does not contribute to the radiation of shear waves, the damping force for this case will be half of that of the sphere embedded in a medium. Hence, the force component related to the damping of the oscillations of the sphere due to the radiation of shear waves for the sphere at the viscoelastic medium interface can be obtained as $F_r=\frac{1}{2}\left(-j\omega\sqrt{\frac{\rho}{G}}R\right)1.5f_0^{1/3}\left(\frac{4E^*\sqrt{R}}{3}\right)^{2/3}U$. Now, the equation for the sphere at an elastic interface in frequency domain is obtained as follows:

$$1.5(-jf_0/\omega)\left(e^{j\omega\tau}-1\right)=\frac{1}{3}\pi R^3(4\rho_s+\rho)(-\omega^2U)+1.5f_0^{1/3}\left(\frac{4E^*\sqrt{R}}{3}\right)^{2/3}\left(1-\frac{1}{2}j\omega\sqrt{\frac{\rho}{G}}R\right)U \tag{8}$$

It should be noted that, the reduced Young's modulus becomes $E^*=E/(1-\nu^2)$ when the sphere is not deformable. The elasticity modulus is related to the shear modulus by $E=2G(1+\nu)$ for homogeneous isotropic materials. Hence, we obtain $E^*=2G(1+\nu)/(1-\nu^2)$ for a homogeneous isotropic material and a rigid sphere. Considering the effect of the viscosity of the medium (η), that is, by replacing G with $(G - j\omega\eta)$, as similar to other studies [24,25], equation (8) can be rewritten as follows:

$$(-jf_0/\omega)\left(e^{j\omega\tau} - 1\right) = \frac{2}{9}\pi R^3(4\rho_s + \rho)(-\omega^2 U) + f_0^{1/3}\left[\frac{8(G-j\eta\omega)(1+v)\sqrt{R}}{3(1-v^2)}\right]^{2/3}\left(1 - \frac{1}{2}j\omega\sqrt{\frac{\rho}{G-j\eta\omega}}R\right)U \tag{9}$$

Hence, we obtained a very useful expression that contains the correct displacement-force relationship for the sphere at a medium interface for dynamic loading as well as the damping term for the oscillations of the sphere at an interface due to the radiation of shear waves. Overall, the dynamic response of the sphere at the viscoelastic interface is obtained as follows:

$$U = \frac{(-jf_0/\omega)\left(e^{j\omega\tau}-1\right)}{-\frac{2}{9}\pi R^3(4\rho_s+\rho)\omega^2 + f_0^{1/3}\left[\frac{8(G-j\eta\omega)(1+v)\sqrt{R}}{3(1-v^2)}\right]^{2/3}\left(1-\frac{1}{2}j\omega\sqrt{\frac{\rho}{G-j\eta\omega}}R\right)} \tag{10}$$

As explained before, the time-domain response of the sphere at the viscoelastic interface can be found by the inverse Fourier transform as:

$$u(t) = \frac{1}{2\pi}\int_{-\infty}^{\infty} \frac{(-jf_0/\omega)\left(e^{j\omega\tau}-1\right)e^{-j\omega t}}{-\frac{2}{9}\pi R^3(4\rho_s+\rho)\omega^2 + f_0^{1/3}\left[\frac{8(G-j\omega\eta)(1+v)\sqrt{R}}{3(1-v^2)}\right]^{2/3}\left(1-\frac{1}{2}j\omega\sqrt{\frac{\rho}{G-j\omega\eta}}R\right)} d\omega \tag{11}$$

3. Results and discussion

The excitation duration τ was divided into N (e.g., 1000) points and the calculations were repeated over the entire time period of interest using the Matlab software (Mathworks, Natick, MA). The displacements of the sphere at the viscoelastic medium interface for different force levels and the excitation time of $\tau = 0.67$ ms are shown in Figure 2a. Here, the properties of medium are $G = 1300$ Pa, $\rho = 1000$ kg/m^3, $\eta = 0.1$ Pa s and $v = 0.45$ and the radius and density of sphere are $R = 0.5$ mm and $\rho_s = 3980$ kg/m^3. In addition, the displacements of the same sphere embedded in the same viscoelastic medium are shown in Figure 2b. In addition to the time-domain data, the spectrums for both cases are shown in Figures 2c and d. As expected, the sphere displacement increases as the force level increases for both models. It is clearly seen that the frequency (or period) of oscillations depends on the amplitude of the excitation force for the model of the sphere at the viscoelastic medium interface; the frequencies of oscillations are 120, 170 and 240 Hz for $f_0 = 0.023$, 0.23 and 2.3 mN, respectively. These results show that the proposed model has the ability to estimate the stiffness change due to the amplitude of the excitation force. As the amplitude of the excitation force increases, the stiffness of the system increases, therefore the frequency of oscillations increases. On the other hand, the frequency of oscillations does not change with force level for the model of the sphere in a medium, since the system's stiffness parameters do not change with the force level for this model. It is seen that there is a peak at 190 Hz for the model of the sphere in a medium for all force levels. Energy concentrates around natural frequencies for all force levels and, as the force level increases, the energy concentration moves to higher frequencies for the interface model. On the other hand, although the energy level increases with force amplitude, its distribution as a function of frequency is the same for all force levels for the embedded model.

The displacements of the sphere for the interface and embedded models and their spectrums for different densities of sphere are shown in Figure 3. It is seen that the frequency of oscillations decreases with the increase of the sphere density for the interface model; the frequencies of oscillations are 310, 240 and 180 Hz for $\rho_s = 1990, 3980$ and 7960 kg/m^3, respectively. The amplitude of the resonance peak decreases and the peak becomes wider as the sphere density decreases for the interface model. It should be noted that, for a viscously damped system, the damping ratio is expressed by $\zeta = \frac{c}{c_{cr}} = \frac{c}{2\sqrt{km}}$. Therefore, the system's damping ratio increases as the density of the sphere (or the effective mass of the system) decreases and this results in a decrease in the displacement amplitude as seen in Figure 3, though the medium viscosity is the same for all three curves in Figure 3.

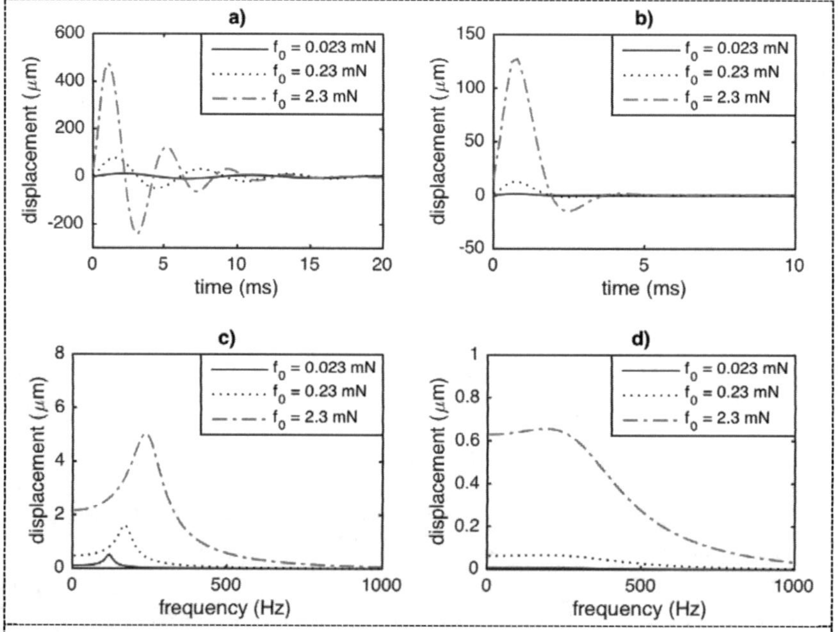

Figure 2. Sphere displacements predicted using the interface (a) and embedded (b) models for different force levels and the excitation time of $\tau = 0.67$ ms and the spectrums obtained using the interface (c) and embedded (d) models ($G = 1300$ Pa, $\rho = 1000$ kg/m^3, $\eta = 0.1$ Pa s, $v = 0.45$, $R = 0.5$ mm, $\rho_s = 3980$ kg/m^3).

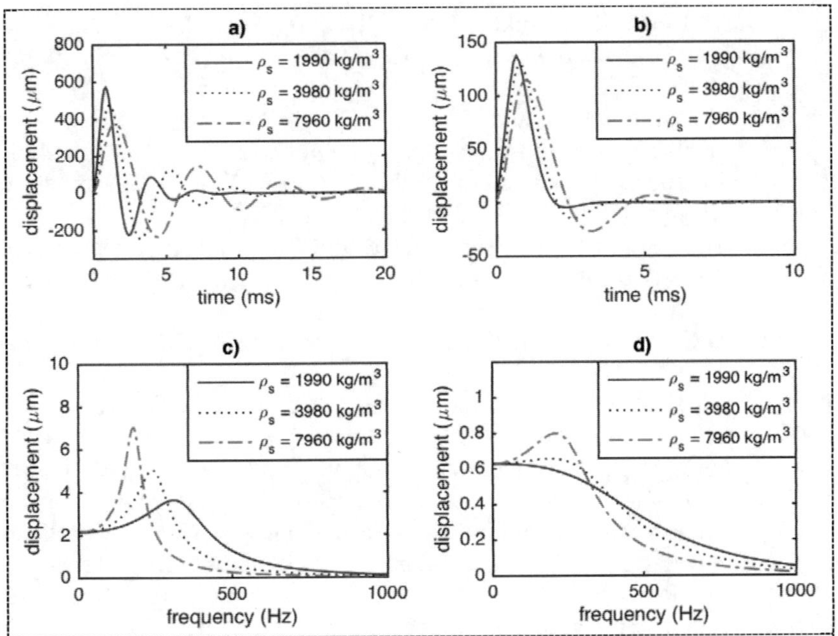

Figure 3. Sphere displacements predicted using the interface (a) and embedded (b) models for the different densities of the sphere and the spectrums obtained using the interface (c) and embedded (d) models ($G = 1300$ Pa, $\rho = 1000$ kg/m^3, $\eta = 0.1$ Pa s, $v = 0.45$, $R = 0.5$ mm, $f_0 = 2.3$ mN, $\tau = 0.67$ ms).

The proposed model can be used in various applications. When the displacements of a sphere with known physical properties at a viscoelastic medium interface to a given excitation force are measured, the model can be used to determine the shear modulus and viscosity of the medium. The model can be used to identify the physical properties of spherical inhomogeneities at viscoelastic interfaces with known mechanical properties by using the measured displacements of the inhomogeneities to a given excitation force. It should be noted that there are some studies for determining the effects of surface elasticity and surface stress by measuring the shifts of resonant frequencies of a beam [29]. The model in this study can provide a new experimental approach for determining the surface elasticity based on the shift of the resonance frequency of the sphere at viscoelastic interface or by using the measured displacements of the sphere with known physical properties at the viscoelastic medium interface to a given excitation force. In some applications, it is difficult, or not possible, to determine the force, such as the acoustic radiation force on spherical inhomogeneities in ultrasound applications. The model proposed in this study can be used to identify the unknown force by measuring the displacements of the sphere with known physical properties at the viscoelastic medium interface with known mechanical properties or the resonance shifts for different force levels.

References

[1] Chen S, Fatemi M and Greenleaf J F. Remote measurement of material properties from radiation force induced vibration of an embedded sphere. *The Journal of the Acoustical Society of America*, 884–889, 112 (3), 2002.

[2] Karpiouk A B, Aglyamov S R, Ilinskii Y A, Zabolotskaya E A and Emelianov S Y. Assessment of shear modulus of tissue using ultrasound radiation force acting on a spherical acoustic inhomogeneity. *IEEE Transactions on Ultrasonics, Ferroelectrics, and Frequency Control*, 2380–2387, 56 (11), 2009.

[3] Argatov I and Mishuris G. *Indentation Testing of Biological Materials* (Springer), 2018.

[4] Erpelding T N, Hollman K W and O'Donnell M. Bubble-based acoustic radiation force elasticity imaging. *IEEE Transactions on Ultrasonics, Ferroelectrics, and Frequency Control*, 971–979, 52 (6), 2005.
[5] Pouliopoulos A N, Burgess M T and Konofagou E E. Pulse inversion enhances the passive mapping of microbubble-based ultrasound therapy. *Applied Physics Letters,* 44102, 113, 2018.
[6] Garcia R and San Paulo A. Dynamics of a vibrating tip near or in intermittent contact with a surface. *Physical Review B*, R13381-R13384, 61 (20), 2000.
[7] Dimitriadis E K, Horkay F, Maresca J, Kachar B and Chadwick R S. Determination of elastic moduli of thin layers of soft material using the atomic force microscope. *Biophysical Journal,* 2798–2810, 82 (5), 2002.
[8] Tan E P S and Lim C T. Nanoindentation study of nanofibers. *Applied Physics Letters,* 123106, 87 (12), 2005.
[9] Koruk H and Choi J J. Displacement of a bubble by acoustic radiation force into a fluid–tissue interface. *The Journal of the Acoustical Society of America*, 2535–2540, 143 (4), 2018.
[10] Koruk H and Choi, J J. Displacement of a bubble located at a fluid-viscoelastic medium interface. *The Journal of the Acoustical Society of America*, EL410–EL416, 145 (5), 2019.
[11] Koruk H, El Ghamrawy A, Pouliopoulos A N and Choi J J. Acoustic particle palpation for measuring tissue elasticity. *Applied Physics Letters,* 223701, 107 (22), 2015.
[12] Bezer J H, Koruk H, Rowlands C J and Choi J J. Elastic deformation of soft tissue-mimicking materials using a single microbubble and acoustic radiation force. *Ultrasound in Medicine & Biology*, 3327-3338, 46 (12), 2020.
[13] Johnson K L. *Contact Mechanics* (Cambridge University Press), 1985.
[14] Darling E M, Zauscher S, Block J A and Guilak F. A thin-layer model for viscoelastic, stress-relaxation testing of cells using atomic force microscopy: Do cell properties reflect metastatic potential? *Biophysical Journal*, 1784–1791, 92 (5), 2007.
[15] Ketene A N, Roberts P C, Shea A A, Schmelz E M and Agah M. Actin filaments play a primary role for structural integrity and viscoelastic response in cells. *Integrative Biology*, 540–549, 4 (5), 2012.
[16] Nayak P R. Contact vibrations. *Journal of Sound and Vibration*, 297–322, 22 (3), 1972.
[17] Tian J, Ogi H and Hirao M. Dynamic-contact stiffness at the interface between a vibrating rigid sphere and a semi-infinite viscoelastic solid. *IEEE Transactions on Ultrasonics, Ferroelectrics, and Frequency Control*, 1557-1563, 51 (11), 2004.
[18] Cheng L, Xia X, Scriven L E and Gerberich W W. Spherical-tip indentation of viscoelastic material. *Mechanics of Materials,* 213–226, 37 (1), 2005.
[19] Qiang B, Greenleaf J, Oyen M and Zhang X. Estimating material elasticity by spherical indentation load-relaxation tests on viscoelastic samples of finite thickness. *IEEE Transactions on Ultrasonics, Ferroelectrics, and Frequency Control*, 1418–1429, 58 (7), 2011.
[20] Yang Y, Zeng Q and Wan L. Dynamic response analysis of the vertical elastic impact of the spherical rock on the metal plate. *International Journal of Solids and Structures*, 287–302, 158, 2019.
[21] Zhou Y. Modeling of softsphere normal collisions with characteristic of coefficient of restitution dependent on impact velocity. *Theoretical & Applied Mechanics Letters*, 021003, 3, 2013.
[22] Sherif H A and Almufadi F A. Identification of contact parameters from elastic-plastic impact of hard sphere and elastic half space. *Wear*, 358–367, 368–369, 2016.
[23] Mijailovic A S, Qing B, Fortunato D and Vliet K J V. Characterizing viscoelastic mechanical properties of highly compliant polymers and biological tissues using impact indentation. *Acta Biomaterialia*, 388–397, 71, 2018.
[24] Ilinskii Y A, Meegan G D, Zabolotskaya E A and Emelianov S Y. Gas bubble and solid sphere motion in elastic media in response to acoustic radiation force. *The Journal of the Acoustical Society of America*, 2338–2346, 117 (4), 2005.

IOP Conf. Series: Materials Science and Engineering 1150 (2021) 012015 doi:10.1088/1757-899X/1150/1/012015

[25] Aglyamov S R, Karpiouk A B, Ilinskii Y A, Zabolotskaya E A and Emelianov S Y. Motion of a solid sphere in a viscoelastic medium in response to applied acoustic radiation force: Theoretical analysis and experimental verification. *The Journal of the Acoustical Society of America*, 1927-1936, 122 (4), 2007.

[26] Laverty Jr S M. *Experimental Hydrodynamics of Spherical Projectiles Impacting On a Free Surface Using High Speed Imaging Techniques* (Massachusetts Institute of Technology), 2004.

[27] Hulme A. A ring-source/integral-equation method for the calculation of hydrodynamic forces exerted on floating bodies of revolution. *Journal of Fluid Mechanics*, 387–412, 128, 1983.

[28] Miloh T. On the oblique water-entry problem of a rigid sphere. *Journal of Engineering Mathematics*, 77–92, 25, 1991.

[29] Zhang Y, Zhuo L J and Zhao H S. Determining the effects of surface elasticity and surface stress by measuring the shifts of resonant frequencies. *Proceedings of The Royal Society A*, 20130449, 469, 2013.

IOP Conf. Series: Materials Science and Engineering 1150 (2021) 012016 doi:10.1088/1757-899X/1150/1/012016

Assessment of the models for predicting the responses of spherical objects in viscoelastic mediums and at viscoelastic interfaces

H Koruk[1]

[1]Mechanical Engineering Department, MEF University, 34396 Istanbul, Turkey

E-mail: korukh@mef.edu.tr

Abstract. Spherical objects, such as bubbles and spheres, embedded in mediums and at viscoelastic interfaces are encountered in many applications, including the determination of material properties. This paper assesses the models for predicting the responses of spherical objects in viscoelastic mediums and at viscoelastic interfaces used in various applications. The models are presented very compactly, and evaluations are performed based on the analyses of the models for the spherical objects in viscoelastic mediums and at viscoelastic interfaces. First, the models for predicting the static displacements of spherical objects are presented and assessed. After that, the models for predicting the dynamic responses of spherical objects are presented and their dynamic behaviours are compared. Then, the models for the deformation of the medium around spherical objects and stress distribution are presented and evaluated. The models and evaluations presented in this study can be exploited in various applications, including biomedical applications.

1. Introduction

Spherical objects, such as bubbles and spheres, embedded in mediums and at viscoelastic interfaces are encountered in many applications [1-4]. For example, bubbles and spheres embedded in a medium have been used to determine material properties, such as the Young's modulus [1,5]. Microbubbles at interfaces are used in therapeutic and diagnostic ultrasound applications [6,7]. Similarly, spheres at interfaces are exploited in atomic force microscopy or indentation tests [8-10]. There have been many studies in recent years to understand the responses of spherical objects in mediums and at viscoelastic interfaces, and to reveal their applications [5,7,11-17]. However, the models for predicting the responses of spherical objects in viscoelastic mediums and at viscoelastic interfaces have not been compared and evaluated in the literature so far.

The models for predicting the responses of bubbles and spheres in viscoelastic mediums and at viscoelastic interfaces are presented and evaluated in this paper. Here, the models are presented very compactly, and evaluations are performed based on the analyses of the models for the spherical objects in viscoelastic mediums and at viscoelastic interfaces. The spherical object with a radius R embedded in a viscoelastic medium and at a viscoelastic interface with a shear modulus G, density ρ and viscosity η is illustrated in Figure 1. Here, first, the models for predicting the static displacements of spherical objects are presented and assessed. After that, the models for predicting the dynamic responses of spherical objects are presented and their dynamic behaviours are compared. Then, the models for predicting the deformation of the medium around the spherical objects and stress distribution are presented and evaluated.

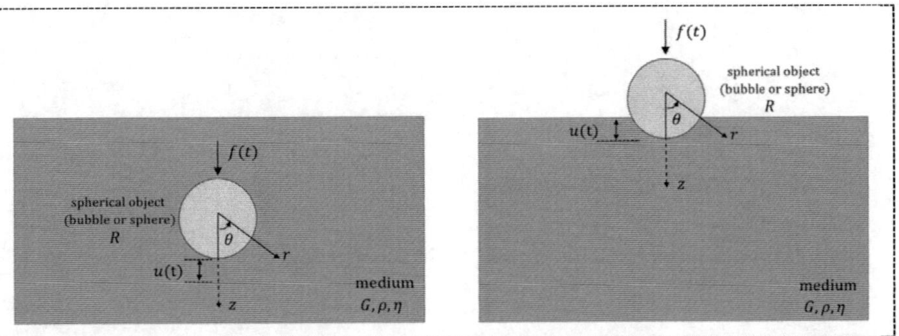

Figure 1. The spherical object embedded in a viscoelastic medium (left) and at a viscoelastic interface (right).

2. Static displacements of the spherical objects

(1) The displacement of the bubble embedded in a soft medium for a static force is given by [11]:

$$u = \frac{f_0}{4\pi G R} \tag{1}$$

(2) The displacement of the bubble at a soft interface for a static force is defined by [14]:

$$u = \frac{f_0}{2\pi G R\left[1-\left(1-\frac{u}{R}\right)^3\right]} \tag{2}$$

(3) The displacement of the sphere embedded in a soft medium for a static force is given by [11]:

$$u = \frac{f_0}{6\pi G R} \tag{3}$$

(4) The displacement of the sphere at an interface for a static force is given by [18]:

$$u = \left(\frac{3f_0}{4E^*\sqrt{R}}\right)^{2/3} \tag{4}$$

Here, f_0 is the force amplitude, G is the shear modulus of the medium, E^* is the reduced Young's modulus, computed as $1/E^* = \left(1 - v_{\text{sphere}}^2\right)/E_{\text{sphere}} + (1 - v^2)/E$ and R is the radius of the spherical object. E and v are the Young's modulus and Poisson's ratio of the medium, respectively. For a rigid sphere, the reduced Young's modulus becomes $E^* = E/(1 - v^2)$. The elasticity modulus is related to the shear modulus by $E = 2G(1 + v)$ for homogeneous isotropic materials.

The stiffness for the four models presented above is plotted as a function of the displacement of the spherical objects in Figure 2. Here, the stiffness is normalized by GR (i.e., $\frac{f_0}{uGR}$) and the displacement is normalized by R (i.e., $\frac{u}{R}$). It is seen that the stiffness of the bubble and sphere embedded in the medium are constant. On the other hand, the stiffness changes nonlinearly with the displacement for the spherical objects at interfaces. The stiffness of the sphere in the medium, the bubble in the medium, and the bubble at the interface is approximately 4.2, 2.8, and 1.4 times, respectively, greater than the stiffness of the sphere at the interface for a displacement equal to the radius of the spherical objects (i.e., $u = R$).

The normalized force (i.e., $\frac{f_0}{GR}$) required for a specific normalized displacement for the four systems presented above is plotted in Figure 3. It is seen that the force required for a specific displacement is highest for the sphere embedded in the medium and lowest for the sphere at the interface. The force required for a specific displacement for the bubble embedded in the medium is higher than the one for the bubble at the interface.

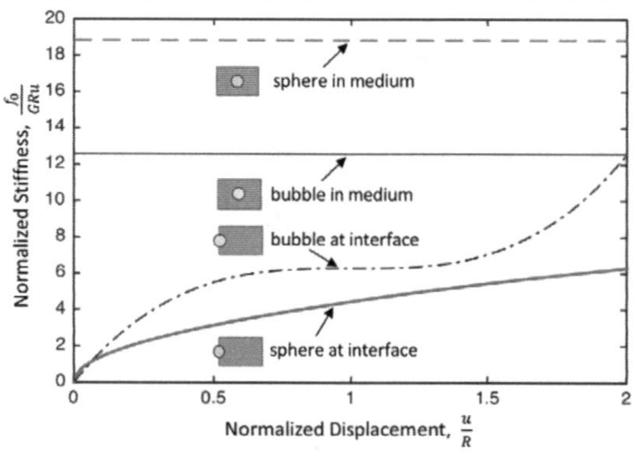

Figure 2. The normalized stiffness as a function of normalized displacement.

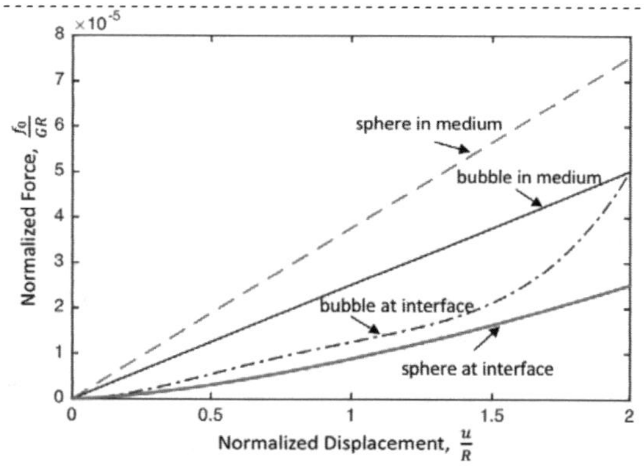

Figure 3. The normalized force as a function of normalized displacement.

3. Dynamic responses of the spherical objects

(i) The frequency-domain equation that couples the displacement of the bubble embedded in a soft viscoelastic medium for a dynamic force is [11]:

$$4\pi(G - j\omega\eta)RU\left(1 - jkR - \frac{1}{6}k^2R^2 + \frac{1}{18}jk^3R^3\right) = F \tag{5}$$

where $k = \frac{\omega}{\sqrt{(G-j\omega\eta)/\rho}}$ is the complex wave number of the shear wave with the frequency ω, ρ and η are the density and viscosity of the medium, $j = \sqrt{-1}$, F is the Fourier transform of the external force $f(t)$, and U is a spectral component of the bubble displacement. Let's now consider the external force as a rectangular pulse with an amplitude of f_0 and a duration of τ (i.e., the constant force f_0 is applied for a short time τ and then removed). It should be remembered that the response to the rectangular pulse simulates the impulse response for small τ values and the step response for large τ values with $0 \leq t \leq \tau$. It can be calculated that the Fourier transform of the rectangular pulse is $F = -\frac{jf_0}{\omega}\left(e^{j\omega\tau} - 1\right)$. Hence, the dynamic displacement of the bubble in the viscoelastic medium can be determined using:

$$u(t) = -\frac{jf_0}{24\pi^2 R}\int_{-\infty}^{\infty}\frac{(e^{j\omega\tau}-1)(3-jkR)e^{-j\omega t}}{\omega(G-j\omega\eta)\left(1-jkR-\frac{1}{6}k^2R^2+\frac{1}{18}jk^3R^3\right)}\mathrm{d}\omega \tag{6}$$

(ii) The frequency-domain equation that couples the displacement of the bubble at a soft viscoelastic medium interface for a dynamic force is given by [14,15]:

$$2\pi(G - j\omega\eta)RU(1 - cos^3\theta_u)\left(1 - jkR - \frac{1}{6}k^2R^2 + \frac{1}{18}jk^3R^3\right) = F \tag{7}$$

where θ_u is the angle corresponding to the displacement $u(t)$. The dynamic displacement of the bubble at the viscoelastic medium interface can be determined using:

$$u(t) = -\frac{jf_0}{12\pi^2 R\left\{1-\left[1-\frac{u(t)}{R}\right]^3\right\}}\int_{-\infty}^{\infty}\frac{(e^{j\omega\tau}-1)(3-jkR)e^{-j\omega t}}{\omega(G-j\omega\eta)\left(1-jkR-\frac{1}{6}k^2R^2+\frac{1}{18}jk^3R^3\right)}\mathrm{d}\omega \tag{8}$$

(iii) The frequency-domain equation that couples the displacement of the sphere embedded in a soft viscoelastic medium for a dynamic force is given as [12]:

$$-m_s\omega^2 U + 6\pi(G - j\omega\eta)RU\left(1 - jkR - \frac{1}{9}k^2R^2\right) = F \quad (9)$$

Here $m_s = \frac{4}{3}\pi R^3\rho_s$ is the mass of the sphere where ρ_s is the density of the sphere. The dynamic displacement of the sphere in the viscoelastic medium can be determined using:

$$u(t) = -\frac{jf_0}{12\pi^2 R}\int_{-\infty}^{\infty}\frac{(e^{j\omega\tau}-1)e^{-j\omega t}}{\omega\left[-\frac{2}{9}R^2\rho_s\omega^2+(G-j\omega\eta)\left(1-jkR-\frac{1}{9}k^2R^2\right)\right]}d\omega \quad (10)$$

(iv) The frequency-domain equation that couples the displacement of the sphere at a viscoelastic medium interface for a dynamic force can be expressed as follows [17]:

$$-\frac{2}{9}\pi R^3(4\rho_s + \rho)\omega^2 U + f_0^{1/3}\left[\frac{8(G-j\omega\eta)(1+\nu)\sqrt{R}}{3(1-\nu^2)}\right]^{2/3}\left(1 - \frac{1}{2}j\omega\sqrt{\frac{\rho}{G-j\omega\eta}}R\right)U = F \quad (11)$$

The dynamic displacement of the sphere at the viscoelastic medium interface can be determined using:

$$u(t) = \frac{1}{2\pi}\int_{-\infty}^{\infty}\frac{(-jf_0/\omega)(e^{j\omega\tau}-1)}{-\frac{2}{9}\pi R^3(4\rho_s+\rho)\omega^2+f_0^{1/3}\left[\frac{8(G-j\omega\eta)(1+\nu)\sqrt{R}}{3(1-\nu^2)}\right]^{2/3}\left(1-\frac{1}{2}j\omega\sqrt{\frac{\rho}{G-j\omega\eta}}R\right)}d\omega \quad (12)$$

The dynamic responses of spherical objects with a radius of $R = 2$ μm in a viscoelastic medium and at a viscoelastic interface ($G = 2000$ Pa, $\rho = 1000$ kg/m^3, $\eta = 0.2$ Pa s, $\nu = 0.45$) for a force of $f_0 = 0.015$ μN and two different durations are plotted in Figure 4. Here, the time duration is divided into N points (e.g., 500), and the calculations are repeated over the entire period of interest using Matlab (MathWorks, Natick, MA, USA) to find the dynamic responses of spherical objects. It is seen that the rate of relaxation is lowest for the bubble at the interface. A force of $f_0 = 0.015$ μN (or 15 nN) produces a steady-state displacement of approximately 0.20, 0.28, 0.78, and 1.13 μm for the sphere in the medium, the bubble in the medium, the bubble at the interface, and the sphere at the interface, respectively, for a radius of $R = 2$ μm.

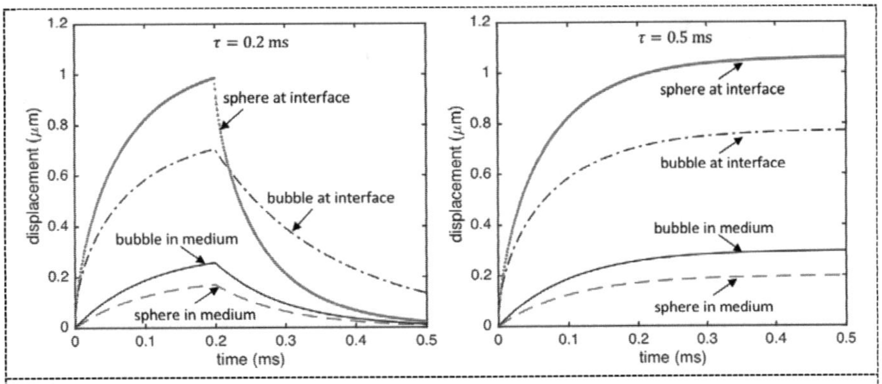

Figure 4. Displacements of spherical objects (G = 2000 Pa, ρ = 1000 kg/m^3, η = 0.2 Pa s, ν = 0.45, R = 2 μm, ρ_s = 3000 kg/m^3, f_0 = 0.015 μN).

The dynamic responses of spherical objects with a radius of $R = 100$ μm in a viscoelastic medium and at a viscoelastic interface ($G = 2000$ Pa, $\rho = 1000$ kg/m^3, $\nu = 0.45$) with two different viscosity values and a force of $f_0 = 30$ μN are plotted in Figure 5. It is seen that, for all the four cases, only the sphere at the interface oscillates (when $\eta = 0.2$ Pa s). However, the sphere does not have any oscillations when the viscosity is increased ($\eta = 0.5$ Pa s). It is seen that the viscosity does not change the steady-state displacements of the spherical objects. The results show that the force producing a displacement of $u = R$ is tens of nano-Newton for spherical objects with a diameter of a few micro meters. The corresponding force is tens of micro-Newton for spherical objects with a diameter of hundred micro meters.

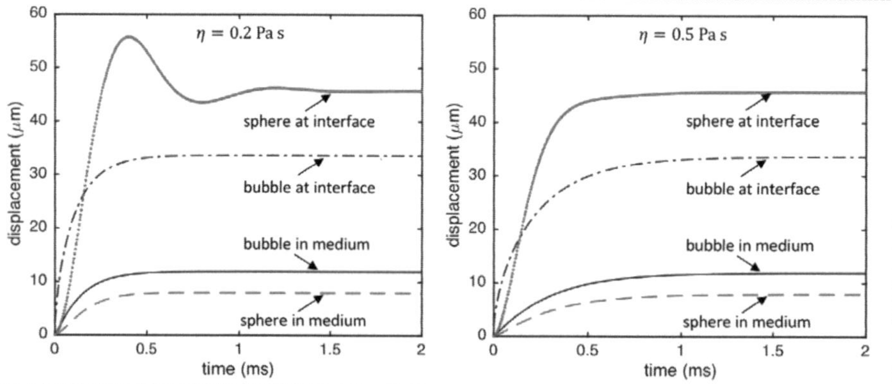

Figure 5. Displacements of spherical objects ($G = 2000$ Pa, $\rho = 1000$ kg/m3, $v = 0.45$, $R = 100$ µm, $\rho_s = 3000$ kg/m³, $f_0 = 30$ µN).

4. Deformation of the medium around spherical objects and stress distribution

(a) Using the formulations by Ilinskii et al. [11], the deformation of the medium and the nonzero stress component for the bubble embedded in a soft medium can be written as:

$$u_r(r,\theta,t) = \frac{R}{r}u(t)\cos(\theta) \tag{13a}$$

$$u_\theta(r,\theta,t) = -\frac{R}{2r}u(t)\sin(\theta) \tag{13b}$$

$$\sigma_r(r,\theta,t) = -\frac{2GR}{r^2}u(t)\cos(\theta) \tag{14}$$

where $u(t)$ is given by equations (1) and (6) for static and dynamic loading, respectively. (b) Using the formulations by Koruk and Choi [14], the deformation of the medium and the nonzero stress component for the bubble at a soft interface can be written as:

$$u_r(r,\theta,t) = \frac{R}{r}u(t)\cos(\theta) \tag{15a}$$

$$u_\theta(r,\theta,t) = -\frac{R}{2r}u(t)\sin(\theta) \tag{15b}$$

$$\sigma_r(r,\theta,t) = -\frac{2GR}{r^2}u(t)\cos(\theta) \tag{16}$$

where $u(t)$ is given by equations (2) and (8) for static and dynamic loading, respectively. (c) Using the formulations by Ilinskii et al. [11], the deformation of the medium and the nonzero stress component for the sphere embedded in a soft medium can be written as:

$$u_r(r,\theta,t) = u(t)\left(\frac{3R}{2r} - \frac{R^3}{2r^3}\right)\cos(\theta) \tag{17a}$$

$$u_\theta(r,\theta,t) = -u(t)\left(\frac{3R}{4r} + \frac{R^3}{4r^3}\right)\sin(\theta) \tag{17b}$$

$$\sigma_r(r,\theta,t) = -\frac{3G}{r}u(t)\left(\frac{R}{r} - \frac{R^3}{r^3}\right)\cos(\theta) \tag{18a}$$

$$\sigma_\theta(r,\theta,t) = \frac{3G}{2r^4}u(t)R^3\sin(\theta) \tag{18b}$$

where $u(t)$ is given by equations (3) and (10) for static and dynamic loading, respectively. (d) The radius of the contact area caused by the sphere at an interface and the maximum pressure within the contact area can be written as:

$$a = \sqrt{Ru(t)} \tag{19}$$

$$p_{\max} = \frac{3f_0}{2\pi Ru(t)} \tag{20}$$

where $u(t)$ is given by equations (4) and (12) for static and dynamic loading, respectively. The stresses underneath the contact area are:

$$\sigma_x = \sigma_y = -p_{\max}\left\{\left[1 - \left|\frac{z}{a}\right| \operatorname{atan}\frac{1}{|z/a|}\right](1+v) - \frac{1}{2(1+z^2/a^2)}\right\} \tag{21a}$$

$$\sigma_z = -\frac{p_{\max}}{1+z^2/a^2} \tag{21b}$$

The stress developed in the viscoelastic medium (G = 2000 Pa, ρ = 1000 kg/m^3, v = 0.45) underneath the spherical object with a radius of R = 2 μm as a function of position for a force of f_0 = 0.015 μN is plotted in Figure 6. Here, the position is measured from the tip contact point and normalized by the radius (i.e., $\frac{p}{R}$). It is seen that the stress is maximum at the tip contact point (p = 0) for the bubble while it is maximum at p/R = 0.4 for the sphere. For the same force level, the magnitude of the maximum stress is approximately 150, 600, 1550, and 1850 Pa for the sphere in the medium, the bubble in the medium, the bubble at the interface, and the sphere in the interface, respectively. As expected, the stress approaches zero far from the contact point for all cases (being negligible for p/R > 8).

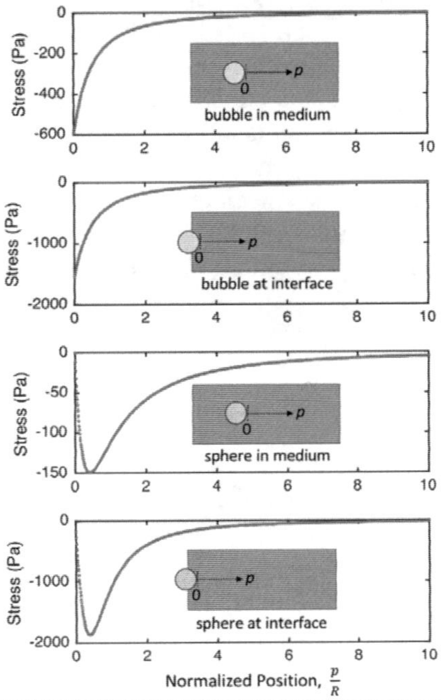

Figure 6. Stress distribution underneath the tip contact point (G = 2000 Pa, ρ = 1000 kg/m^3, v = 0.45, R = 2 μm, f_0 = 0.015 μN).

5. Concluding remarks

In this paper, the models for predicting the responses of spherical objects in viscoelastic mediums and at viscoelastic interfaces are assessed. First, the models for predicting the static displacements of spherical objects are presented and assessed. After that, the models for predicting the dynamic responses of spherical objects are presented and their dynamic behaviors are compared. Then, the models for predicting the deformation of the medium around the spherical objects and stress distribution are presented and evaluated.

- The force increases linearly with the displacement of the bubble and sphere embedded in a medium, though it changes nonlinearly with the displacement of the bubble and sphere at an interface. The force needed for a displacement equal to the radius of the spherical object for the sphere in the medium, the bubble in the medium, and the bubble at the interface is approximately 4.2, 2.8, and 1.4 times, respectively, greater than that of the sphere at the interface. The force producing a steady-state displacement equal to the radius of the spherical object is tens of nano-Newton for spherical objects with a diameter of a few micro meters (for a medium shear modulus of a few thousand Pascal). The corresponding force is tens of micro-Newton for spherical objects with a diameter of hundred micro meters.
- The rate of relaxation is low for the bubble at the interface, compared to the sphere at the interface and the bubble and sphere in the medium. For the same radius and material properties, although there are no oscillations for all other three systems, the sphere at the interface can oscillate. However, the sphere does not have any oscillations when its radius is small and the medium viscosity is high. The viscosity does not change the steady-state displacements of the spherical objects.
- The stress is maximum at the tip contact point for the bubble, while it is maximum at the point where the ratio of position to radius is 0.4 for the sphere. For the same level of force, the magnitude of the maximum stress for the bubble in the medium, the bubble at the interface, and the sphere in the interface is approximately 4, 10, and 12 times, respectively, greater than that for the sphere in the medium. For all four cases, stress is negligible when the ratio of position to radius is greater than 8.

The models and evaluations presented in this study can be used in many practical applications. For example, they can be used to design experiments and to identify the elastic and damping properties of materials at the nano- and micro-scale in biomedical applications. They can be exploited to determine the responses of spherical objects in viscoelastic mediums and at viscoelastic interfaces, and to identify the stresses developed in mediums around the spherical objects in practical applications. The experimental evaluation of the four cases presented in this paper is considered as the topic of a future work.

References

[1] Chen S, Fatemi M and Greenleaf J F. Remote measurement of material properties from radiation force induced vibration of an embedded sphere. *The Journal of the Acoustical Society of America*, 884–889, 112 (3), 2002.

[2] Rigaud E and Perret-Liaudet J. Experiments and numerical results on non-linear vibrations of an impacting Hertzian contact. Part 1: harmonic excitation. *Journal of Sound and Vibration*, 289-307, 265 (2), 2003.

[3] Cheng L, Xia X, Scriven L E and Gerberich W W. Spherical-tip indentation of viscoelastic material. *Mechanics of Materials,* 213–226, 37 (1), 2005.

[4] Karpiouk A B, Aglyamov S R, Ilinskii Y A, Zabolotskaya E A and Emelianov S Y. Assessment of shear modulus of tissue using ultrasound radiation force acting on a spherical acoustic inhomogeneity. *IEEE Transactions on Ultrasonics, Ferroelectrics, and Frequency Control*, 2380–2387, 56 (11), 2009.

[5] Erpelding T N, Hollman K W and O'Donnell M. Bubble-based acoustic radiation force elasticity imaging. *IEEE Transactions on Ultrasonics, Ferroelectrics, and Frequency Control*, 971–979, 52 (6), 2005.

[6] Acconcia C, Leung B Y C, Hynynen K and Goertz D E. Interactions between ultrasound stimulated microbubbles and fibrin clots. *Applied Physics Letters,* 53701, 103 (5), 2013.
[7] Pouliopoulos A N, Burgess M T and Konofagou E E. Pulse inversion enhances the passive mapping of microbubble-based ultrasound therapy. *Applied Physics Letters,* 44102, 113, 2018.
[8] Garcia R and San Paulo A. Dynamics of a vibrating tip near or in intermittent contact with a surface. *Physical Review B*, R13381-R13384, 61 (20), 2000.
[9] Dimitriadis E K, Horkay F, Maresca J, Kachar B and Chadwick R S. Determination of elastic moduli of thin layers of soft material using the atomic force microscope. *Biophysical Journal,* 2798–2810, 82 (5), 2002.
[10] Tan E P S and Lim C T. Nanoindentation study of nanofibers. *Applied Physics Letters,* 123106, 87 (12), 2005.
[11] Ilinskii Y A, Meegan G D, Zabolotskaya E A and Emelianov S Y. Gas bubble and solid sphere motion in elastic media in response to acoustic radiation force. *The Journal of the Acoustical Society of America*, 2338–2346, 117 (4), 2005.
[12] Aglyamov S R, Karpiouk A B, Ilinskii Y A, Zabolotskaya E A and Emelianov S Y. Motion of a solid sphere in a viscoelastic medium in response to applied acoustic radiation force: Theoretical analysis and experimental verification. *The Journal of the Acoustical Society of America*, 1927-1936, 122 (4), 2007.
[13] Koruk H, El Ghamrawy A, Pouliopoulos A N and Choi J J. Acoustic particle palpation for measuring tissue elasticity. *Applied Physics Letters,* 223701, 107 (22), 2015.
[14] Koruk H and Choi J J. Displacement of a bubble by acoustic radiation force into a fluid–tissue interface. *The Journal of the Acoustical Society of America*, 2535–2540, 143 (4), 2018.
[15] Koruk H and Choi, J J. Displacement of a bubble located at a fluid-viscoelastic medium interface. *The Journal of the Acoustical Society of America*, EL410–EL416, 145 (5), 2019.
[16] Bezer J H, Koruk H, Rowlands C J and Choi J J. Elastic deformation of soft tissue-mimicking materials using a single microbubble and acoustic radiation force. *Ultrasound in Medicine & Biology*, 3327-3338, 46 (12), 2020.
[17] Koruk H. Development of a model for predicting dynamic response of a sphere at viscoelastic interface: A dynamic Hertz model. *IOP Conference Series: Materials Science and Engineering*, 2021.
[18] Johnson K L. *Contact Mechanics* (Cambridge University Press), 1985.

IOP Conf. Series: Materials Science and Engineering 1150 (2021) 012017 doi:10.1088/1757-899X/1150/1/012017

Nanoencapsulation of ethylvanillin using Electrohydrodynamic technology

Megdi Eltayeb

Mechanical Engineering Department, College of Engineering, Prince Mohammad Bin Fahd University, KSA

Email: meltayeb@pmu.edu.sa

Abstract. An electrohydrodynamic method for preparation of ethylvanillin-loaded polymer-lipid-based nanoparticles. Various parameters of electrohydrodynamic are optimized for generating ethylvanillin-loaded nanoparticles by controllable electrohydrodynamic parameters. Ethylvanillin-loaded polymer-lipid-based nanoparticles with mean diameters of around 64-90nm with encapsulation efficiency in the range 67-71% are generated with all the flow rate from 10 to 25μl/min, the stearic acid-ethylcellulose-ethylvanillin concentration ratio of 4:1:1.6 wt%, and ethylvanillin content of 0.16%. The size of the ethylvanillin-loaded polymer-lipid-based nanoparticles production is examined using scanning electron microscopy.

1. Introduction

Currently, the food industry is confronted with the challenges of developing and applying novel technology that can generate high-quality and healthy foods. Meeting these complex engineering and scientific challenges, novel technologies are instigated for developing new processes, products, and tools in this industry. Encapsulation of active ingredient (flavor) in different polymer compositions and sizes include in ranging from micro to nano-size is widely used in many industries like; pharmaceutical, food, and agriculture [1, 2]. The key benefit of encapsulation includes; coating, the barrier between the chemically, environmentally unstable active ingredient and environmental factors like light, O_2 and humidity, etc [3]. It also allows to control the release of active ingredients for delayed or prolong release [4-8]. Further, it can be used to convert liquid active ingredients into a solid (dry) system [9, 10]. The electrohydrodynamic study is a physical process result by the applied voltage force to the material solution or suspension, then the solution formed at the outlet of a tube, form a different mode of the jet (i.e. droplet, microdroplet, cone jet, multi con-jet) [11]. Electrohydrodynamic can be successfully used to produce solid lipid nanoparticles to encapsulate and deliver active compounds. The electrohydrodynamic spraying technique is inexpensive, flexible, simple, and single-step technology [12]. Therefore, it can encapsulate active ingredients in polymeric matrices. The work aimed to prepare solid lipid nanoparticles of ethyl vanillin using the single jet electrohydrodynamic technology. The polymers used to encapsulate ethylvanillin are ethylcellulose and stearic acid. Further optimization of ethylcellulose and stearic acid was also conducted to furnish minimum size particles with monodispersivity and uniform size.

2. Materials and methods

2.1. Materials
All solvent and reagents were used as received without further purification. The regents used are ethylvanillin, ethylcellulose, and stearic acid in powder form, and (95%) ethanol. All the tests were repeated five times.

2.2. Preparation and characterization of ethylvanillin-loaded polymer-lipid-based nanoparticles solution
Stearic acid and ethylcellulose were first dissolved in ethanol in different concentrations (10–50 mg/ml) and a magnetic stirrer was used to dissolve polymers until a clear solution was obtained. Ethylvanillin in various concentrations was also dispersed in the polymeric solution. The concentration of ethyl vanillin was varied in the range of 0 to 5 mg/ml, to prepare particles with a maximum loading of ethylvanillin in terms of weight. The physical properties of the spraying solution were determined.

2.3. Preparation of ethylvanillin-loaded polymer-lipid based nanoparticle
Figure 1 Illustration of electrohydrodynamic spraying setup, used to prepare ethylvanillin-loaded polymer-lipid-based nanoparticle. Spraying solutions (ethylvanillin-loaded polymer-lipid-based nanoparticle) were pumped out, using a mechanical syringe pump from a metal needle (inner diameters of the needle were 450 μm). The spraying solution formed a compound drop under the exit of the needle. The products were collected on a petri-dish containing double distilled water. A camera was used to observed the behavior of spraying solution at the outlet of the needle.

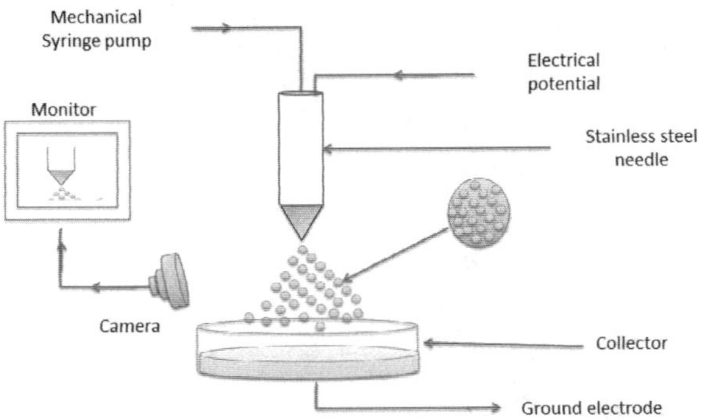

Figure 1. Schematic illustration of the electrohydrodynamic spraying set-up used for ethylvanillin-loaded polymer-lipid-based nanoparticle preparation.

3. Results and Discussion
Spraying is a widely applied encapsulation process in the food industry being flexible and involving less commercial operational cost. This study focuses on the preparation of polymer-lipid-based nanoparticles for the encapsulation of ethylvanillin using electrohydrodynamic technology. Electrohydrodynamic parameters can be controlled with optimization of applied voltage, flow rate. The electrohydrodynamic factors (i.e. applied voltage, flow rate) may influence the polymer-lipid-based nanoparticle encapsulation system and the study also discusses these factors. In this study, the nanoparticles were tremendously affected by varying the composition of ethylvanillin, stearic acid, and ethylcellulose. The effect of stearic acid and ethylcellulose ratio on the ethylvanillin-loaded polymer-lipid-based nanoparticle size was studied by changing the ratio of the stearic acid, and ethylcellulose ratio C from (0- 5% w/v). Figure 1 shows the electrohydrodynamic spraying setup, used to produce ethylvanillin-loaded polymer-lipid-based nanoparticles. Polymer-lipid-based nanoparticle solution with four different concentrations (stearic acid, and ethylcellulose w/w%), were sprayed using electrohydrodynamic spraying (Figure 1) technique. These particles produced by cone-jet are mono-dispersed.

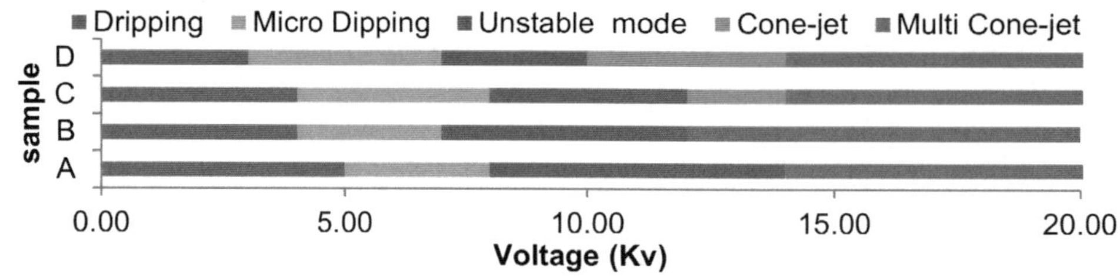

Figure 2. Relationship between applied voltage and polymer Concentrations; (A) 1:4, (B) 2:3, (C) 3:2, and (D) 4:1at 10μl/min flow rate.

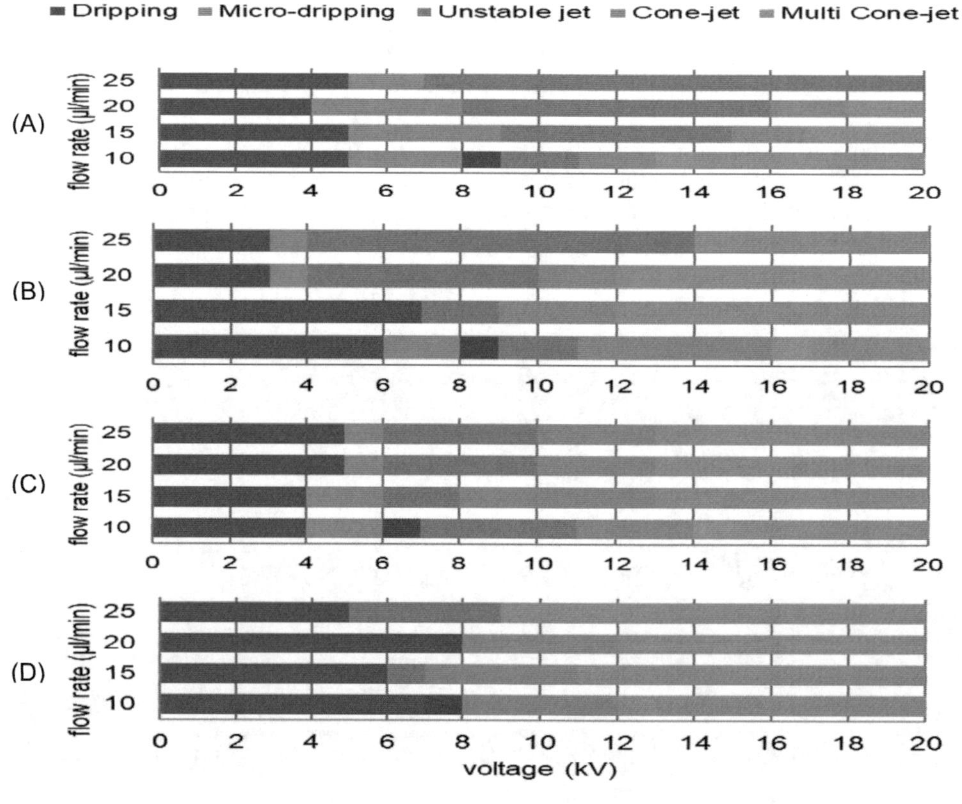

Figure 3. Stearic acid, and ethylcellulose ratio 4:1 effect of varying ethylvanllin concentration (A) 5%; (B) 2.5%; (C) 1.6% and (D) 1.2%, at varying flow rate and applied voltage

The investigation of the relationship between electric voltage and polymeric concentration is shown in (Figure 2). Initially, the voltage from 1 to 20kV was applied in small increments until a cone jet was shaped around (10 - 14kV) for concentrations C and D (i.e. 3:2 and 4:1). A stable cone jet can be obtained when the flow rate was 10μl/minute. The multi-jet mode was detected above 14kV for the entire concentrations and the dripping and micro dripping mode for less than 10kV.
Ethyl vanillin concentrations (1.2, 1.6, 2.5, and 5 w/w%) were added to the selected spraying solution of stearic acid, and ethylcellulose (4:1 w/w%), which was optimized previously to furnish nano-size and uniform particles. Varying voltages (1-20kV) and varying flow rate (10 - 25μl/min) were applied to obtain the stable cone-jet particle [13]. It was found that ethanol was evaporated quickly during the spraying from the metal needle to the grounded collector. The ethylvanillin-loaded polymer-lipid-based nanoparticle acquired structure (Figure 4) is influenced by the evaporation of ethanol from the ethylvanillin-loaded polymer-lipid-based nanoparticle droplet [14].

IOP Conf. Series: Materials Science and Engineering 1150 (2021) 012017 doi:10.1088/1757-899X/1150/1/012017

Experimental results show that the applied electric voltage and flow rate of the sprayed solution can control the size of the produced ethylvanillin-loaded polymer-lipid-based nanoparticle using an electrohydrodynamic setup. Figure 4 shows that ethylvanillin-loaded polymer-lipid-based nanoparticles of various diameter sizes were produced at varying voltage. It is indicated from Figure 4 that different size particles were produced at a different flow rate. This indicates that flow rate has a tremendous effect not only in terms of generating cone-jet but also different size particles. It also confirms that at a lower flow rate small/nanoparticles were generated. However, at a high flow rate comparatively bigger sizes were obtained.

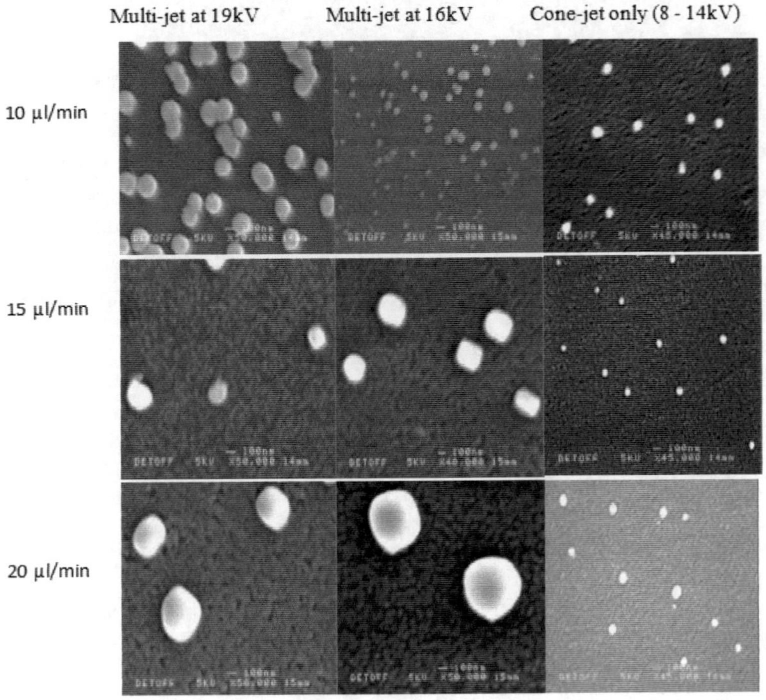

Figure 4. SEM images of ethylvanillin-loaded polymer-lipid-based nanoparticle prepared at 4:1:1.6 concentration of stearic acid, ethylcellulose, and ethyl vanillin, under varying flow rates (10, 15, and 20μl/min) and applied voltage set at cone-jet only (8 - 14kV), and multi cone-jet 16 and 19kV.

Ethanol evaporation rate from the formed ethylvanillin-loaded polymer-lipid-based nanoparticle droplet significantly affects the ethylvanillin-loaded polymer-lipid-based nanoparticle structures. Fast evaporation of the solvent causes cavities inside particles as seen in the ethylvanillin-loaded polymer-lipid-based nanoparticle. UV spectroscopic analysis of nanoparticles shown the entrapment efficiency of optimized ethylvanillin-loaded polymer-lipid-based nanoparticles was as high as possible. These values indicate that the electrohydrodynamic technique was able to generate ethylvanillin loaded into polymer-lipid-based nanoparticles with entrapment efficiencies range between 67 – 71%. The loss of ethylvanillin during the electrohydrodynamic technique was calculated with ethylvanillin-loaded polymer-lipid-based nanoparticles, being in total 30±3% losses. The loss was generally a result of scattering of the ethylvanillin-loaded polymer-lipid-based nanoparticle nanoparticles to the connection tube walls, collector, and during the electrospraying [15].

4. Conclusions

This study has demonstrated that solid lipid nanoparticles entrapping flavor can be generated by simple single needle electrohydrodynamic. Based on the findings of this study the diameter of the solid lipid nanoparticles could be controlled by adjusting the ethylvanillin-stearic acid-ethylcellulose concentration and electrohydrodynamic processing parameters. The nanoparticles were found to have a homogenous structure, with the diameter being dependent upon electrohydrodynamic processing parameters and the concentration of ethylvanillin-stearic acid-ethylcellulose in the electrosprayed solution. The entrapment efficiency was similarly found to be dependent on the ethylvanillin-stearic acid-ethylcellulose concentration ratio. The entrapment efficiency of ethylvanillin into the nanoparticles in the range of 67-71%. The properties of the ethylvanillin remained physically unchanged during the process. The electrohydrodynamic process for the production of stearic acid-ethylcellulose nanoparticles may be a suitable form for the manufacture of stearic acid and ethylcellulose nanoparticles.

References

[1] Jafari, S.M., et al., *Nano spray drying of food ingredients; materials, processing and applications.* Trends in Food Science & Technology, 2021.

[2] Zhong, Q. and L. Zhang, *Nanoparticles fabricated from bulk solid lipids: Preparation, properties, and potential food applications.* Advances in colloid and interface science, 2019. **273**: p. 102033.

[3] Eltayeb, M., et al., *Electrosprayed nanoparticle delivery system for controlled release.* Materials Science and Engineering: C, 2016. **66**: p. 138-146.

[4] Atif, R., I. Yahya, and M. Eltayeb. *Simulation of Drug Release from Biodegradable and Erodible Polymers.* in *2019 International Conference on Computer, Control, Electrical, and Electronics Engineering (ICCCEEE)*. 2019. IEEE.

[5] Yahya, I., et al., *Polymeric Nanoparticles as Drug Delivery Systems for Controlled Release.* Advanced Science, Engineering and Medicine, 2020. **12**(2): p. 263-270.

[6] Ahmed, L., et al., *Study the Using of Nanoparticles as Drug Delivery System Based on Mathematical Models for Controlled Release.* International Journal of Latest Technology in Engineering, Management & Applied Science-IJLTEMAS, 2019. **8**(5): p. 52-56.

[7] Yahya, I., et al., *Mathematical Modeling of Diffusion Controlled Drug Release Profiles from Nanoparticles.* International Journal of Research and Scientific Innovation (IJRSI), 2019. **6**(5): p. 287-291.

[8] Ahmed, L., I. Yahya, and M. Eltayeb. *Computational Modeling of Drug Release Profiles from Swellable Polymeric Nanoparticles.* in *2019 International Conference on Computer, Control, Electrical, and Electronics Engineering (ICCCEEE)*. 2019. IEEE.

[9] Sahani, S. and Y.C. Sharma, *Advancements in applications of nanotechnology in global food industry.* Food Chemistry, 2020: p. 128318.

[10] Yahya, I., et al., *Utilization of solid lipid nanoparticles loaded anticancer agents as drug delivery systems for controlled release.* International Journal of Engineering Applied Sciences and Technology, 2019. **3**(12): p. 7-16.

[11] Eltayeb, M., E. Stride, and M. Edirisinghe, *Electrosprayed core-shell polymer-lipid nanoparticles for active component delivery.* Nanotechnology, 2013. **24**(46).

[12] Coelho, S.C., B.N. Estevinho, and F. Rocha, *Encapsulation in food industry with emerging electrohydrodynamic techniques: Electrospinning and Electrospraying-A review.* Food Chemistry, 2020: p. 127850.

[13] Eltayeb, M., et al., *Preparation of solid lipid nanoparticles containing active compound by electrohydrodynamic spraying.* Food Research International, 2013. **53**(1): p. 88-95.

[14] Eltayeb, M., E. Stride, and M. Edirisinghe, *Electrosprayed core-shell polymer-lipid nanoparticles for active component delivery.* Nanotechnology, 2013. **24**(46): p. 465604.

[15] Eltayeb, M., E. Stride, and M. Edirisinghe, *Preparation, characterization and release kinetics of ethylcellulose nanoparticles encapsulating ethylvanillin as a model functional component.* journal of functional foods, 2015. **14**: p. 726-735.

IOP Conf. Series: Materials Science and Engineering 1150 (2021) 012018 doi:10.1088/1757-899X/1150/1/012018

Catalytic pyrolysis of plastic materials using natural zeolite catalysts synthesized from volcanic ash

María Vargas[1], Danny Tupayachy-Quispe[2], Francine Roudet[3], Sophie Duquesne[4], Jonathan Almirón[5]

[1,2]Laboratorio de Ciencia de los Materiales, Universidad Católica de Santa María, Umacollo, Arequipa, Perú.
[3] Univ. Lille, Institut Mines-Télécom, Univ. Artois, Junia, ULR 4515 - LGCgE – Laboratoire de Génie Civil et géoEnvironnement, F-59000 Lille, France
[4]Laboratoire des Procédés d'Élaboration de Revêtements Fonctionnels, École Nationale Supérieure de Chimie de Lille, Villeneuve-d'Ascq, France
[5]Escuela Profesional de Ingeniería Ambiental, Facultad de Ingeniería de Procesos, Universidad Nacional de San Agustín de Arequipa, Cercado, Arequipa, Perú.

Corresponding author: mvargasv@ucsm.edu.pe

Abstract. Plastic waste and its inadequate final disposal are a serious problem of environmental pollution, which has increased considerably in this time of pandemic due to the use of masks, gloves, food packaging or others, which will eventually reach the oceans, increasing the problem of the islands that are made up of plastic waste. Since the final disposal techniques of plastic waste are insufficient, in this research the catalytic pyrolysis has been evaluated using zeolites synthesized from volcanic ash of the Ubinas volcano (Peru), as catalysts. For this, two catalysts were synthesized, Z1 and Z2 zeolite, made at 150 and 180 °C in a 5 M NaOH solution, respectively. The pyrolysis was carried out at a catalyst/polymer ratio of 10 and 20 % and it was compared with a synthetic zeolite (ZSM-5) and without catalyst at a reaction temperature of 450 °C. The essays were carried out with PP virgin and plastic waste. It was determined that catalytic pyrolysis using natural zeolites synthesized from volcanic ash are appropriate for the treatment of plastic materials and that they tend to generate a higher proportion of pyrolytic gases and do not generate carbonaceous residues.

1. Introduction

Pyrolysis or cracking is a tertiary recycling technique through which organic polymers decompose at high temperatures into components of lower molecular weight in the form of fuel gases, oils and solid residue under conditions of an oxygen-free atmosphere, therefore, dioxin is not produced and carbon monoxide and dioxide emissions are reduced [1], [2], [3]. Currently, the pyrolysis is being studied to be an alternative for the mitigation of urban plastic waste in the environment. There are two types of pyrolysis, the thermal and the catalytic, of them, the catalytic pyrolysis has multiple advantages over thermal since it improves the efficiency of the process by reducing the residence time and affects the selectivity of the pyrolytic products, in addition, the use of catalysts such as zeolites generates high-quality products that do not require subsequent downstream stages [4], [5]. The performance of pyrolytic products will depend on the type of plastic waste used as raw material since this will determine the boiling point of pyrolytic products [4]. The main factors that affect the pyrolysis process are the temperature, the retention time, the composition of the raw material and the catalyst [2], as well as, the type of reactor, pressure, residence time, type of circulating gas and heating rate [6]. The control of these factors will determine the performance and distribution of the products [6]. Despite all the potential advantages of catalytic pyrolysis, the process has some limitations such as the high energy demand that requires, the cost of the catalyst and the low reuse of the catalysts [2]. So, the recommended solutions for these challenges are the search for cheaper catalysts, the regeneration of catalysts and the general optimization of the process.

Catalysts are chemical substances that accelerate chemical reactions, decrease the residence times and the reaction temperature (requiring less energy), and in the case of polymer pyrolysis, they improve the conversion and quality of fuels, increasing their efficiency against the thermal pyrolysis [2], [5], [7]. The characteristics of the catalyst will influence the performance and quality of the pyrolytic products to be obtained [8]. The heterogeneous catalysts are the most widely used because they are easy to separate from pyrolytic products (gases, liquids or solids) and also resist temperatures up to 1300 °C [4]. The main characteristics that are sought in catalysts is that they have the presence of Lewis and Brönsted acid sites since both are involved in the mechanism of polymers degradation. The physical properties of the catalysts are also important, such as the surface area, the pore size and the pore volume [6]. Among the catalysts used for polymer pyrolysis are natural and synthetic zeolites (ZSM-5, HUSY, Hβ and HMOR), silica-alumina, alumina and FCC (Fluid Catalytic Cracking) [2], [4], [6].

Zeolites are crystalline aluminosilicates that, when are dehydrated, they develop a porous structure with ion exchange capacity [6]. Due to their high crystallinity, surface area and selectivity they are very active in catalytic processes. The potential and economic use of natural zeolite was tested for the catalytic pyrolysis of polypropylene, polyethylene, polystyrene and polyethylene terephthalate against synthetic zeolite, this despite of the low surface area of natural zeolite, its microporous structure and low acidity [7]. There have been studies of natural zeolites with a low surface area, where their surfaces were modified by impregnating them with Ni, Ni-Mo, Co and Co-Mo and they observed that the modification improved their catalytic properties by increasing their activity and selectivity [9].

Volcanic ash is a residue that contains silica and alumina, being the ash from the Ubinas volcano rich in silica (45.26%) and alumina (13.40%) [10], the flame-retardant activity of polymeric of this volcanic ash has been studied [11]. So, in the present research it was sought to take advantage of its composition as a source of raw material for the synthesis of natural zeolites, which would lower the cost of the catalysts in the pyrolysis process. Therefore, two types of zeolites were synthesized that were tested in the pyrolytic process for virgin polypropylene (PP) and for plastic waste, comparing it with thermal pyrolysis.

2. Experimental procedure

2.1. Materials

The raw material for the synthesis of natural zeolites was the volcanic ash from the Ubinas volcano located in Moquegua, Peru. The composition of volcanic ash is 45.26% SiO_2, 13.40 % Al_2O_3, 6.72% Fe_2O_3 which has a SiO_2/Al_2O_3 molar ratio of 5.73 [12]. For the comparison of the pyrolysis process, the synthetic zeolite, ZSM-5, of the ZEOLYST INTERNATIONAL brand and CBV 5524G model was used with a composition of 100 % w/w zeolite, SiO_2/Al_2O_3 molar ratio of 50 and a surface area of 425 m^2/g. In the pyrolysis process, an extra pure grade nitrogen gas (99.99%) (PRAXAIR P-4631-H) was used as recirculating gas and the polypropylene PP 5707N from SABIC brand and the collected plastic waste containing polypropylene as the main component were used as polymers.

2.2. Synthesis of natural zeolites

The collected volcanic ash was dried and ground in a mono planetary mill at 450 rpm for 25 min until obtaining a powder with a particle size of 75 μm. Subsequently, 10 g of volcanic ash was placed in a teflon with 60 ml of 1.5 M NaOH and then it was taken to a batch reactor. The batch reactor was placed into the oven at 150 or 180 °C for 12 h, which corresponds to Z1 and Z2, respectively. The sample was left to rest for 2 days at room temperature and then it was washed with 500 ml of distilled water and rest for 1 day for its sedimentation. Then the vacuum filtration of the sample was proceeded to eliminate the greatest amount of water and finally to be dried in the stove at 100 °C for 3 h. The dried sample was ground and placed in a bottle.

2.2.1. Characterization of natural zeolites. Z1 and Z2 natural zeolites were texturally characterized, their surface area (S_{BET}), pore size and pore volume were determined by BET adsorption and desorption isotherms in an ASAP 2020 Micromeritics apparatus with N_2 gas flow at 77 K.

2.3. Pyrolysis process

The natural zeolites and polymers were weighed on a RADWAG brand analytical balance, model AS 220.R2 with readability of 0.1 mg. The catalyst/polymer ratio (% w/w) of 10 and 20 was used for each of the zeolites. Fast pyrolysis was carried out in a NanYang XinYu model SK2-2-12TPB3 tubular furnace with a 60 x 200 mm chamber at 450 °C, with a heating rate of 10 °C/min and with a continuous circulation of N_2 gas. Arriving at the set point temperature, the crucible containing the zeolite and the polymer was introduced, which was kept in the oven for 20 min. The gas generated passed through a cooling trap contained in a thermos with liquid nitrogen where the liquids were recovered, and the gases were received in the extractor hood. After the process, the crucible and the cooling trap were weighed to determine the percentage of yield (wt%) of the solids, liquids and waxes pyrolytic. The pyrolysis was carried out with the virgin polymer (polypropylene) and the zeolite with the highest yield was selected for the pyrolysis of plastic waste. Table 1 shows the coding of the samples used in the pyrolysis process and their conditions.

Table 1. Sample code and pyrolysis process conditions

Sample code	Zeolite type	Polymer type	Catalyst/polymer ratio (% p/p)	Reaction temperature	Residence time	Polymer weight (g)
T	None	PP	-	450	20	1
Z1-C1	Z1	PP	10	450	20	1
Z2-C1	Z2	PP	10	450	20	1
ZSM-5-C1	ZSM-5	PP	10	450	20	1
Z1-C2	Z1	PP	20	450	20	1
Z2-C2	Z2	PP	20	450	20	1
ZSM-5-C2	ZSM-5	PP	20	450	20	1
PW-Z2-C1	Z2	Plastic waste	10	450	20	1

3. Results and discussion

3.1. Characterization of zeolites

Table 2 shows the textural properties of the natural zeolites synthesized from the volcanic ash of the Ubinas volcano.

Table 2. Textural properties of natural zeolites from Ubinas volcano.

Sample	Pore size (Å)	Pore volume (mm^3/g)	Micropore area (m^2/g)	S_{BET} (m^2/g)
Z1	96.84	47.12	8.08	25.91
Z2	146.81	39.70	8.60	11.43

The pore size, shown in table 1, point out the Z1 and Z2 natural zeolites are mesoporous catalysts while the ZSM-5 synthetic zeolite is microporous as indicated by other authors [13], [14]. These characteristics will determine the distribution of the pyrolytic products as indicated in figure 1.

3.2. Yields of pyrolytic products

The pyrolytic products obtained were carbonaceous solids (char), liquids, waxes and gases. Figure 1 shows the performance of solids, liquids and waxes by varying the catalyst/polymer ratio and it is observed that there is a difference between the thermal pyrolysis with the catalytic pyrolysis (using the zeolites synthesized from volcanic ash and with the synthetic zeolite ZSM-5). Being the percentage of the gaseous product higher in the catalytic pyrolysis with a maximum value of 76.42 ± 4.95% with the ZSM-5 zeolite and the minimum value of 47.26 ± 2.30% without the use of a catalyst. Therefore, it could be inferred that the use of zeolites favors a greater generation of gases.

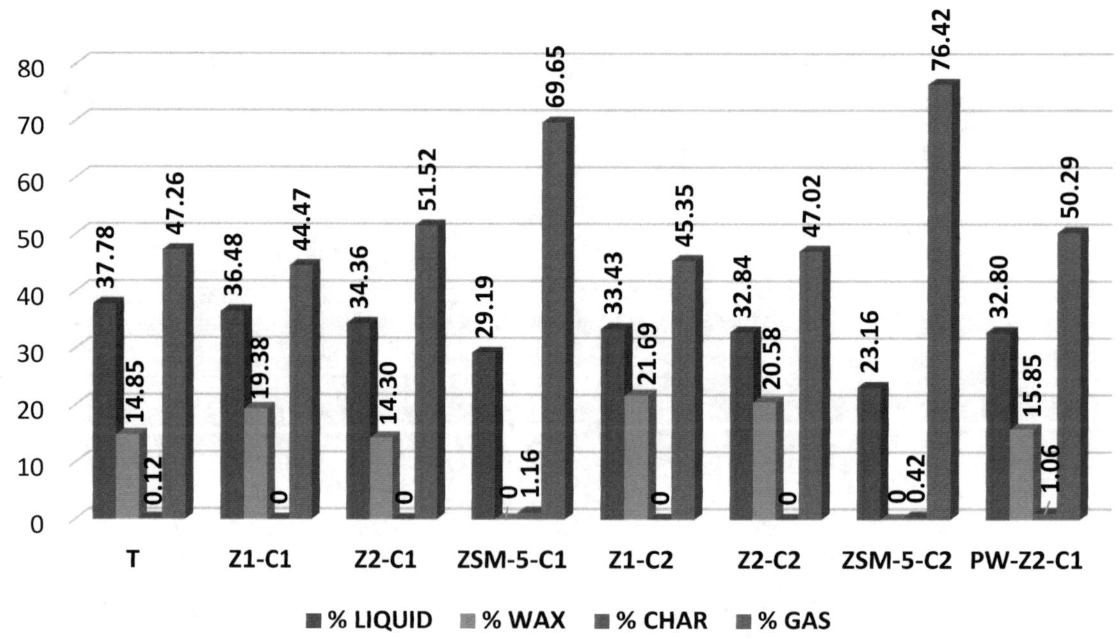

Figure 1. Pyrolytic residues (%) obtained in the catalytic and thermal pyrolysis process using PP and plastic waste.

It is observed that the ZSM-5 zeolite, which is the one with the highest surface area (425 m^2/g), is the one that has produced the greatest amount of gases, this due to the fact that catalysts with a high surface area (S_{BET}) allows a greater contact between the reactants and the catalyst surface, which leads to an increase in the cracking reaction rate, producing more gases than liquids [2]. However, despite the high difference in S_{BET} compared to Z1 and Z2 zeolite, with these ones an increase in the conversion to gases has also been achieved and it is much higher with the Z2 zeolite, which has a larger pore size with a higher micropore area. This is probably because it has zigzagging channels, typical of zeolites, that hinders the formation of larger molecules, such as wax and liquids, within the pores. In addition,

considering that in the catalytic pyrolysis mechanism, in its initial stages thermal cracking occurs on the external surface of the catalyst, where the wax is mainly generated and later, the pores act as selective channels, whose size influences, and where the gases will generate [2].
This would corroborate that the catalytic activity depends on the structural characteristics of the zeolites, among them, the pore size [2], [15], which are influenced by their synthesis conditions. This would explain the reason why there is a lower gas yield with Z1 zeolite whose pore size is smaller than Z2 zeolite. This behavior was also observed by a natural zeolite with a low S_{BET} (3.21 m^2/g), a pore size of 8.34 Å with a microporous structure and low acidity whose pyrolytic yield was 75.4% gas, 14% liquid and 10.7% solid residue. This is probably due to the fact that the catalyst supports the initial degradation of PP, which would increase the generation of gases but not that of liquids [14], [16].

Additionally, a variation is observed between catalytic pyrolysis when treating PP and plastic waste, this occurs because the plastic waste also present other additives aside from the PP that can affect the process. This is the reason why when treating plastic waste, washing and segregation according to their typology, is important [17], [18]. However, in the present study, no major difference in performance has been observed both in obtaining liquids (34.36 ± 3.20% and 32.80 ± 3.29% for Z2C1 and PW-Z2C1), waxes (14.30 ± 5.93% and 15.85 ± 1.55%, for Z2C1 and PW-Z2C1) and gases (51.52 ± 5.32% and 50.29 ± 3.92%, for Z2C1 and PW-Z2C1); while if there was carbonaceous residue generation with the PW-Z2C1 (1.06 ± 0.22%). Moreover, it should be noted that when using plastic waste there was the generation of solid carbonaceous residue while there was no generation when using virgin polypropylene, this due to the presence of other additives in addition to polypropylene. It is worth mentioning that according to Santella *et al.* [15] a small pore size favors the generation of solid waste (coke) which is typical of the ZSM-5 zeolite which in all cases has produced char.

4. Conclusion

It is concluded that the natural zeolites synthesized from the Ubinas volcano have a catalytic activity, in addition the type of zeolite and the catalyst/polymer ratio influence on the performance and distribution of the pyrolytic products. A higher yield of pyrolytic gases was obtained with the catalytic pyrolysis reaching a maximum of 76.42 ± 4.95% (with the ZSM-5 zeolite) compared to 47.26 ± 2.30% (thermal pyrolysis). Between Z1 and Z2 zeolites, a higher gas yield was obtained with the Z2C1 sample (51.52 ± 5.32%). The highest yield of pyrolytic liquids was obtained with thermal pyrolysis (37.78 ± 1.64%). In the case of pyrolytic waxes, the highest yield was obtained with Z1C2 sample with a value of 21.69 ± 2.21%. No solid residues (coke) were obtained with Z1 and Z2 zeolite; however, if was obtained it with ZSM-5 zeolite and with thermal pyrolysis. When using plastic waste and Z2C1 sample, a higher gas yield was obtained (50.29 ± 3.92%).

Reference

[1] Jin Z, Chen D, Yin L, Hu Y, Zhu H. Molten waste plastic pyrolysis in a vertical falling film reactor and the influence of temperature on the pyrolysis products. Chinese Journal of Chemical Engineering. 2018; 26(2): p. 400-406.
[2] Miandad R, Barakat MA, Aburiazaiza AS, Rehan M, Nizami AS. Catalytic Pyrolysis of Plastic Waste: A Review. Process Safety and Environmental Protection. 2016;102: p. 822-838.
[3] Singh R, Ruj B. Time and temperature depended fuel gas generation from pyrolysis of real world municipal plastic waste. Fuel. 2016; 174: p. 164-171.
[4] Al-Salem SM, Antelava A, Constantinou A, Manos G, Dutta A. A review on thermal and catalytic pyrolysis of plastic solid waste (PSW). Journal of Environmental Management. 2017; 197: p. 177-198.
[5] Kunwar B, Cheng HN, Chandrashekaran SR, Sharma BK. Plastics to fuel: a review. Renewable and Sustainable Energy Reviews. 2016; 54: p. 421- 428.
[6] Sharuddin S, Abnisa F, Daud W, Aroua M. A review on pyrolysis of plastic wastes. Energy Conversion and Management. 2016;115: p. 308-326.

[7] Miandad R, Barakat MA, Rehan M, Aburiazaiza AS, Ismail IMI, Nizami AS. Plastic waste to liquid oil through catalytic pyrolysis using natural and synthetic zeolite catalysts. Waste Management. 2017; 69: p. 66-78.

[8] Till Z, Varga T, Sója J, Miskolczi N, Chován. Kinetic identification of plastic waste pyrolysis on zeolite-based catalysts. Energy Conversion and Management. 2018; 173: p. 320-330.

[9] Sriningsih W, Saerodji MG, Trisunaryanti W, Triyono , Armunanto R, Falah II. Fuel Production from LDPE Plastic Waste over Natural Zeolite Supported Ni, Ni-Mo, Co and Co-Mo Metals. Procedia Environmental Sciences. 2014; 20: p. 215-224.

[10] Vargas M, Hermoza-Gutierrez M, Tupayachy-Quispe DP, Almirón J, Huanca Z. PK, Velasco-Lopez FJ. Manufacture of geopolymeric mortars from ash coming from the ubinas volcano, assessment of its mechanical, physical and microstructural properties. Revista Boliviana de Química. 2020; 37(3): p. 148-159.

[11] Almirón J, Roudet F, Duquesne S. Influence of volcanic ash, rice husk ash, and solid residue of catalytic pyrolysis on the flame-retardant properties of polypropylene composites. Journal of Fire Sciences. 2019; 37(4-6): p. 434-451.

[12] Calderón N, Vargas M, Almirón J, Bautista A, Velasco F, Tupayachy-Quispe D. Influence of the Activating Solution and Aggregates in the Physical and Mechanical Properties of Volcanic Ash Based Geopolymer Mortars. IOP Conference Series: Materials Science and Engineering. 2021; 1054(012003): p. 1-9.

[13] Ratnasari DK, Nahil MA, Willi PT. Catalytic Pyrolysis of Waste Plastics using Staged Catalysis for Production of Gasoline Range Hydrocarbon Oils. Journal of Analytical and Applied Pyrolysis. 2017; 124: p. 631-637.

[14] Miandad R, Barakat MA, Aburiazaiza AS, Rehan M, I.M.I. I, Nizami AS. Effect of plastic waste types on pyrolysis liquid oil. International Biodeterioration & Biodegradation. 2017; 119: p. 239-252.

[15] Santella C, Cafiero L, De Angelis D, La Marca F, Tuffi R, Vecchio Ciprioti S. Thermal and catalytic pyrolysis of a mixture of plastics from small waste electrical and electronic equipment (WEEE). Waste Management. 2016; 54: p. 143-152.

[16] Almeida D, Marques M. Thermal and catalytic pyrolisis of plastic waste. Polímeros: Ciência e Tecnologia. 2016; 26(1): p. 44-51

[17] Wong SL, Ngadia N, Abdullah TAT, Inuwa IM. Current state and future prospects of plastic waste as source of fuel: A review. Renewable and Sustainable Energy Reviews. 2015; 50: p. 1167–1180.

[18] Sogancioglu M, Ahmetli G, Yel E. A Comparative Study on Waste Plastics Pyrolysis Liquid Products Quantity and Energy Recovery Potential. Energy Procedia. 2017; 118: p. 221-226.

Acknowledgments
To FONDECYT-CONCYTEC for the financial support and to the National School of Chemistry of the University of Lille in France for its support.

Estimating the flexural strength of concrete using compressive strength as input value in a deep learning model

José A. Guzmán-Torres[1], Francisco J. Domínguez-Mota[1] and Elia M. Alonso-Guzmán[1].

[1]Civil Engineering Faculty, UMSNH, Ciudad Universitaria, 58004 Morelia, Michoacán, México.

E-mail: jaguzman@umich.mx.

Abstract. The flexural strength is a critical parameter for designing many concrete structures such as rigid pavements, beams, and bridges. The standard test for concrete is the compressive strength due to its ease of implementation. There are many proposed methods for estimating flexural strength values with enough accuracy, although it is necessary to enhance the accuracy for this estimation, and this research suggests the use of artificial intelligence methods to accomplish this goal. Artificial Intelligence has been one of the most efficient approaches for estimating material parameters because of its efficient performance. This research presents the development of a data-driven Deep Neural Network for predicting the flexural strength in concrete based on just the compressive strength test. The proposed model analyses a concrete mixture with starch and a fluidizer. The model employs a Rectified Linear Unit function and a Sigmoid function in its architecture as activation functions and a considerable perceptron's number. Results from the analysis show an excellent accuracy of over 90%, which is remarkable. This approach showed satisfactory performance in flexural strength prediction for the analysed concrete mixture.

1. Introduction

Currently, there are plenty of methods to estimate the performance of concrete structures. One approach that appears from the numerical methods is artificial intelligence. Artificial intelligence can be defined as a computer's ability to solve a problem without being explicitly programmed for that task. It is known that concrete is the most traditional material used in construction because of its broad versatility. Also, this material can be malleable compared to other materials [1]. Hence, it is necessary to have alternatives to quantify the resistance of concrete elements efficiently, that is, to obtain the most suitable characterization in the least possible time.

There are many Machine Learning applications in the construction field as the prediction of the tensile strength just with measurements of electrical resistivity, as reported in [2], or the detection of damage in high structures detailed in [3].

The rupture modulus, also known as flexural strength, is an essential value used to determine the behaviour of concrete structures such as beams, bridges, and rigid pavements. Obtaining this value may sometimes be a difficult task because the amount of material needed to perform this test is considerable; is costly, time-consuming, and requires special testing equipment. In some countries such as Mexico, this test is omitted and replaced by the compressive strength test. Under this consideration, the constructors prefer the use of the following equation:

$$Fs = k * Cs^{1/2} \quad (1)$$

Where k is a constant value that depends on concrete properties and is estimated arbitrarily in some cases, Cs represents the compressive strength of concrete, and Fs is the flexural strength. This paper proposes a deep neural network (DNN) model suitable to predict the flexural strength (Fs) of concrete specimens. These kinds of techniques are used in forecasting tasks, particularly in this material [4]. The compressive strength (Cs), for short, is used as an input parameter in this research, and the Fs is the outcome or also known as the target value. The concrete mixture used for this study was analysed and characterized previously in a laboratory under controlled conditions. Before the modelling process, the laboratory results were examined, and we found the correlation between the features involved in this study. In order to determine the correlation between both input and target parameters, we use the determination coefficient r, also known as Pearson's correlation factor. The details of the developed model along with the validation are provided in the following sections.

2. Research significance

This paper proposes the architecture of a DNN to predict the flexural strength value only with compressive strength as a predictive variable in concrete materials. Further, a unique architecture is proposed for achieving this aim where the fine-tuning of hyperparameters for the best performance of the neural network was tested by inspection. This paper generates a robust model with the potential to analyse different data and get an accuracy close to the initial model.

3. Methodology

3.1. Concrete Mixture

For this study, the concrete mixture analysed contains starch and a fluidizer as additives. The design water-cement rate was equal to 0.35, and the mixture design was following the ACI method [5]. The cement employed in this research corresponds to CPC 40 R, which according to the Mexican cement classification standards [6], is a composite Portland cement with a compressive strength resistance of 40 MPa. The materials employed in the concrete mixtures belong to a bank placed in Morelia, Michoacán, México. The aggregates were both crushed and volcanic. The compressive strength was the critical design parameter in the concrete mixtures. The mixture was monitored in different stages; 40, 60, 90, and 120 days. In Table 1, material properties for this study are shown.

Table 1. Material properties.

Type of aggregates	Properties	Crushed aggregates	Volcanic aggregates
Fine aggregates (Sand)	Percent Absorption	-	2.75
	Density (g/ml)	-	2.59
	Apparent specific gravity(g/cm^3)	-	1.42
	Percent Surface Moisture	-	0.19
	Percent Mesh #200	-	5.45
Coarse aggregates (Gravel)	Percent Absorption	3.40	-
	Density (g/ml)	2.80	-
	Apparent specific gravity(g/cm^3)	1.51	-
	Percent Surface Moisture	1.82	-

3.2. Data collection

First, 390 specimens were tested and monitored, 195 for Cs and 195 for Ts. Nevertheless, we removed the outliers using the standard deviation criterium, which considers the Z-score value. After the cleaning data, the final dataset keeps 183 data for each group. By convention, we segmented the dataset into two groups: training and test, and 70% and 30%, were considered respectively. In this research, the uncertainty of data is ignored because this assumption is commonly accepted in the literature and how is detailed in [7, 8].

3.3. Deep Neural network development

In the model, each layer is assigned with the value of a variable $a_{0,}a_{1,}a_2, \ldots, a_n$. The term a comes from the word axion. The axions are stated for the number of layers contained in the neural network. Each vector is known as an activation vector and is equal in size to the number of neurons that make up the previous layer. Vector a_0 looks like:

$$a_0 = \begin{Bmatrix} a_{0(0)} \\ a_{0(1)} \\ \vdots \\ a_{0(n)} \end{Bmatrix}. \tag{2}$$

The weights generated in each connection are indicated with the w variable. These match the values of the weights that define the input of a neuron. w is known as a matrix that extends to the order of the neuron's number between the layers it is detected. This definition can be generalized by expression (3)

$$[w_{ij}] = \begin{bmatrix} w_{00} & \cdots & w_{0j} \\ \vdots & \ddots & \vdots \\ w_{i0} & \cdots & w_{ij} \end{bmatrix}. \tag{3}$$

Additionally, the sum of a neuron in each layer known as the bias is performed. This variable allows to regulate the threshold of the neuron's behaviour and is given by the expression (4)

$$b_0 = \begin{Bmatrix} b_0 w_{b0(0)} \\ b_0 w_{b0(1)} \\ \vdots \\ b_0 w_{b0(n)} \end{Bmatrix}. \tag{4}$$

Each neuron receives the sum of the products made between the weights and the input data. Thus, it is possible to generalize the equation of a neuronal network as

$$\sum_{i=0}^{n} f\left(w_{ij}a_i + b_j\right) \tag{5}$$

and, for this study, Equation (5) can be rewritten as

$$k_n = \sum_{i=0}^{2} f\left(w_{ij}a_i + b_j\right) \tag{6}$$

where a_i corresponds to the input vector, k_n is the output or the final result of the model, w_{ij} is the weight matrix, b_j is the bias vector, and f represents the activation function. The artificial neuron is considered as a linear function with an adaptable weight matrix. The first four hidden layers operate with ReLU activation functions, and the final hidden layer works with a Sigmoid activation function.

The DNN implementation was done in Phyton's environment with frameworks such as TensorFlow. The values contained in the dataset were normalized to minimize the errors and minimize the computation time. The DNN is composed of four hidden layers with 200 neurons for each hidden layer. ReLU activation breaks the linearity on the hidden layers, and by inspection, we found that this activation performs adequately in this model. The final hidden layers worked with a Sigmoid activation function. The DNN contains six layers, an input layer, four hidden layers, and an output layer. The DNN presented in this study has 200 neurons in each hidden layer.

3.4. Evaluation methods

The accuracy was assessed by comparing the actual value y against the predicted ($y_{predicted}$). The metrics used for evaluating the DNN performance, take into consideration the weights and the loss function and were root mean square error (RMSE), mean absolute error (MAE), and mean squared error (MSE) as follows:

$$\text{RMSE} = \left(\frac{1}{n}\sum_{i=0}^{n}\left(y_i - y_{predicted}\right)^2\right)^{1/2} \quad (7)$$

$$\text{MAE} = \frac{1}{n}\sum_{i=0}^{n}\left|y_i - y_{predicted}\right| \quad (8)$$

$$\text{MSE} = \frac{1}{n}\sum_{i=0}^{n}\left(y_i - y_{predicted}\right) \quad (9)$$

4. Results and discussions

We observe the correlation value between the predictor variables, the histograms, and the diagram distribution of each test in Figure (1), and the correlation value is equal to 0.81. This correlation fluctuates within -1 and +1 with 0, meaning no correlation. Correlations of -1 or +1 imply an exact linear relation, according to [9].

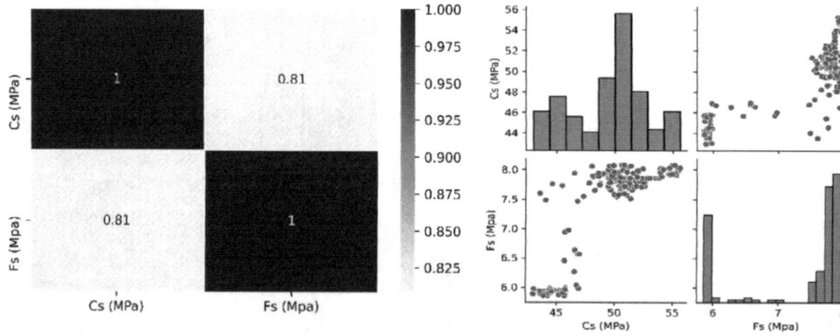

(a) (b)

Figure 1. Data Analysis outcomes. (a) Correlation matrix (b) Histograms and distribution of dataset.

In the modelling process, 121,201 parameters were calculated by the model. The DNN uses a regularization method called early stopping, and this hyperparameter is equal to 3 to avoid overfitting problems. The model is trained during 350 epochs using the Adam function as an optimizer with a modest learning rate of 0.001 and a batch size of 1. Each hidden layer contains 200 neurons, also known as perceptrons. Of note is that whether the number of layers is increasing, the performance model does not enhance. The accuracy in the training and test stage reaches 76.78% and 92.32%, respectively. The loss function performance in the training and test process is shown in Figure (2). The performance expected through the training and test loss considers that the losses decrease as the number of epochs increases. Finally, the values for RMSE, MAE, and MSE are equal to 0.0848, 0.0596, and 0.0072, respectively.

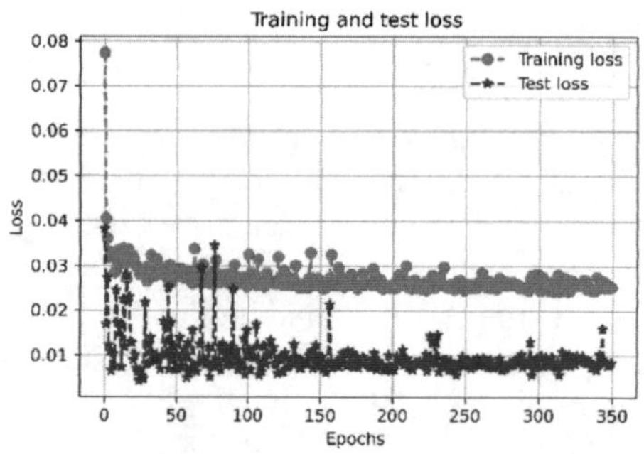

Figure 2. Loss performance over the number of epochs.

In this case, we can see that the model generally achieves a good fit, with train and test learning curves converging.

5. Conclusions

In this paper, a deep neural network was modelled with the capability to predict and compute the flexural strength value of concrete based on a compressive strength test with an accuracy of almost 92% in a validation dataset. This study proposed a different approach to estimate the flexural strength, which considers algorithm implementation for achieving competent results. Sigmoid and ReLU activation functions showed to reach higher accuracy compared to an individual activation function use. Thus, the combination of different activation functions had significantly improved the performance of the DNN for this study. The model architecture is determined through extensive hyperparameter fine-tuning.

This analysis and modelling can be extended to estimate other concrete features, such as tensile strength, ultrasonic pulse velocity, electrical resistivity, and other properties. This methodology might improve the optimization of the concrete mixture design due to it is possible to achieve great results with a low computational cost. It is desirable to study the application of DNN for different problems in future concrete research. Future work will focus on improving the performance of the model by increasing the performance as well as the applicability of the model to other parameters.

The approach used for the presented model in this paper achieves a satisfactory accuracy and is desirable for its use instead of the application of a theoretical equation that might not achieve the best results and depends on another unknown parameter as k value.

References

[1] Torres-Acosta, A. A., & Díaz-Cruz, L. A. (2020). Concrete durability enhancement from nopal (opuntia ficus-indica) additions. *Construction and Building Materials*, *243*, 118170.

[2] Guzmán Torres, J. A., Domínguez Mota, F. J., Alonso-Guzmán, E. M., Martínez-Molina, W., Tinoco Ruiz, J. G., Chavez-Garcia, H. L., ... & Arreola Sánchez, M. (2020). Prediction of the Tensile Strength and Electrical Resistivity of Concrete with Organic Polymer and their Influence on Carbonation Using Data Science and a Machine Learning Technique. In Key Engineering Materials (Vol. 862, pp. 72-77). Trans Tech Publications Ltd.

[3] Rafiei, M. H., & Adeli, H. (2017). A novel machine learning-based algorithm to detect damage in high-rise building structures. *The Structural Design of Tall and Special Buildings*, *26*(18), e1400.

[4] Nguyen, T., Kashani, A., Ngo, T., & Bordas, S. (2019). Deep neural network with high-order neuron for the prediction of foamed concrete strength. Computer-Aided Civil and Infrastructure Engineering, 34(4), 316-332.

[5] Dixon, D. E., Prestrera, J. R., Burg, G. R., Chairman, S. A., Abdun-Nur, E. A., Barton, S. G., ... & Carter, A. C. (1991). Standard Practice for Selecting Proportions for Normal, Heavyweight, and Mass Concrete (ACI 211.1-91).

[6] Organismo Nacional de Normalización y Certificación de la construcción y Edificación, S. C. (2004). NMX-C-414-ONNCCE-2004: Industria de la contrucción – cementos hidráulicos – especificaciones y métodos de prueba. ONNCCE, S. C.

[7] Yaseen, Z. M., Deo, R. C., Hilal, A., Abd, A. M., Bueno, L. C., Salcedo-Sanz, S., & Nehdi, M. L. (2018). Predicting compressive strength of lightweight foamed concrete using extreme learning machine model. Advances in Engineering Software, 115, 112-125.

[8] Antoniades, A., Spyrou, L., Martin-Lopez, D., Valentin, A., Alarcon, G., Sanei, S., & Took, C. C. (2018). Deep neural architectures for mapping scalp to intracranial EEG. International journal of neural systems, 28(08), 1850009.

[9] Bonamente, M. (2017). Statistics and analysis of scientific data. Springer, 117.

Acknowledgments

The authors wish to thank the materials laboratory Ing. Luis Silva Ruelas for providing the necessary equipment to obtain data. Thanks to the CIC and Aula-CIMNE for funding this research.

IOP Conf. Series: Materials Science and Engineering 1150 (2021) 012020 doi:10.1088/1757-899X/1150/1/012020

Numerical Analysis of Initially Imperfect Pretensioned Bolted Connections

Mostafa Nour Eldin M. Abdallah[1], Emad Salem[2], Sherif M. Ibrahim[3], Abdelrahim K.Dessouki[4]

[1] Ph.D. Candidate, Structural engineering Department, Ain Shams University, Cairo, Egypt
[2] Professor, Civil engineering Department, Al-Azhar University, Cairo, Egypt
[3] Professor, Structural engineering Department, Ain Shams University, Cairo, Egypt
[4] Professor, Structural engineering Department, Ain Shams University, Cairo, Egypt

E-mail: g16021527@eng.asu.edu.eg

Abstract. Bolted steel end-plate rigid connections are used to transfer straining actions between different parts of steel structures. However, during structural assembly some of these connections are erected with initial imperfections in form of gap between connected parts. This paper investigates numerically initially imperfect end plate connections subjected to major external bending moment and major shearing force. A finite element model (FEM) is developed to represent the behavior of these initially imperfect connections. The presented FEM is verified against experimental testing presented by the authors in previous work. The effect of the gap angle between connected parts on the connection bending strength and moment-curvature relationship is investigated by varying its value from 1.5° to 6°. The results of initially imperfect connections are compared to those of perfect ones and it is concluded that the loss in bending strength of imperfect connections varies from 11.5 % to 22 % for gap angles from 1.5° to 6° respectively.

1. Introduction

Threaded connections, in particular bolted joints, are a common engineering feature found in most equipments and structures. In preloaded joints the bolts are first tightened sufficiently to establish closure of the joint with alignment of the mating components producing the required bolt preload and (more importantly) a compressive load at the faying surface. The faying surface is the joints prepared contact face. Preloading is intended to maintain closure of the joint and make it perform as a single continuous member.

It has been observed that in many cases the end plates cannot come to full contact although bolts are fully tightened to their required pretension value, this imperfection can either be in joint's tension side or compression side, loaded or unloaded structures (load independent). Fig. (1) shows connections with initial imperfection in existing structures, an illustration of initially imperfect connections is presented in Fig. (2).

Figure 1. Connections with initial imperfections from existing structures

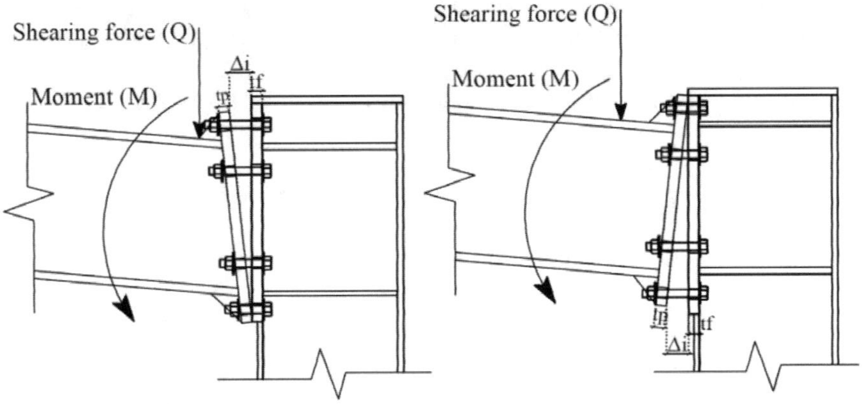

Figure 2. Connections with initial imperfections

Many research projects were carried out in the past few decades to classify the connection types [1 and 2], to develop the mathematical model for connection stiffness [3], and to study the influence of joint flexibility on the behaviour of steel framed structures [4]. However rather less work has been done to evaluate the initial imperfection of the end-plate that is observed after construction and causes a separation of the connection sides at the contact interface on the behaviour of the connections. Based on the finite element analysis, Wheeler [5] studied the influence of the initial imperfection of the end-plate on the behaviour of the tubular moment end-plate connections. and found that the initial stiffness of the connections tended to decrease by introduction of imperfection in the end-plates.

In 2007, Chen and Du [6] studied the residual deformation of welding and its effect on the capacity and behaviour of the bolted end plate. They performed an experimental work on 21 specimens of which 3 were control ideal specimens while 6 specimens were with V-shaped initial imperfection with various degrees, 6 specimens with W-shaped initial imperfection with various degrees and 6 specimens with C-shaped initial imperfection with various degrees, with a total of 21 specimens as shown in Fig. (3):

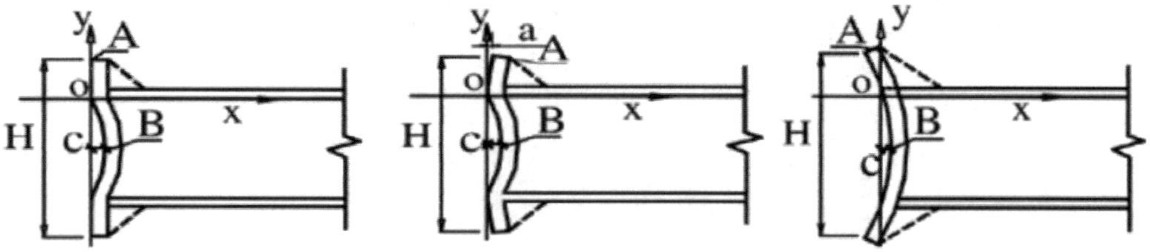

Figure 3. V shape, W shape and C shape imperfection respectively as studied by [6]

It was concluded from their work that in case of imperfection, limited to the criterion that depth of imperfection is less than or equal H/300, gives 4.7% to 8.5% reduction in the rotational stiffness in case of V-shape, while in W-shape about 18% to 19% reduction in the rotation stiffness was observed.

Duana et al. [7] studied the initial imperfection of bolted end-plate connection where the thickness of end-plate and column flange can't meet the code provisions based on a field investigation. The authors performed reliability analysis of steel portal frames, then the Monte Carlo simulation technique was used to estimate the failure probability of steel portal frame system. They concluded that:

1. The assumption of fully rigid connections in the reliability analysis is quite unsafe and can lead to underestimating the failure probability.
2. Insufficient thickness of column flange has a significant effect on safety of portal frame for both flush and extended connections.
3. The insufficient thickness of end-plate has slight effect.

2. Research Significance

For end plate with high strength pretensioned bolts, the current codes and design guidelines [8-10] do not include the case of initially imperfect end plate connections, these codes neither set criteria for accepting or rejecting the initial imperfection as shown in Fig. (1-2), nor state a method of assessing the effect of initial imperfection on the ultimate load capacity of the connection. So, in this paper, a FEM model is developed to study initially imperfect connections and a limited parametric study is also presented to quantify the effect of the gap angle on the bending strength.

3. Finite Element Modelling of Connections

In 2020, Abdallah et.al [11] conducted an experimental program on a group of initially imperfect connections to study their moment deflection behaviour of the initially imperfect connections and bending strength and compare them with ideal connection. They tested four specimens fabricated as a plate girder, namely (I240 BUS of web200x10mm and flanges 120x20mm). The specimens consisted of an ideal connection, and three connections with 2°, 4° and 6° initial imperfection gap angle. Each was loaded up to bolt(s) failure, schematic of the test data are shown in Fig. (4-5).

Bolts used were with nominal diameter of 12mm and nominal grade of 8.8 pretensioned type bolts. The tightening stage was performed with a manual torque wrench to provide a pretension load of 37kN. The steel used for all parts of specimens was S355 (360/520 MPa).

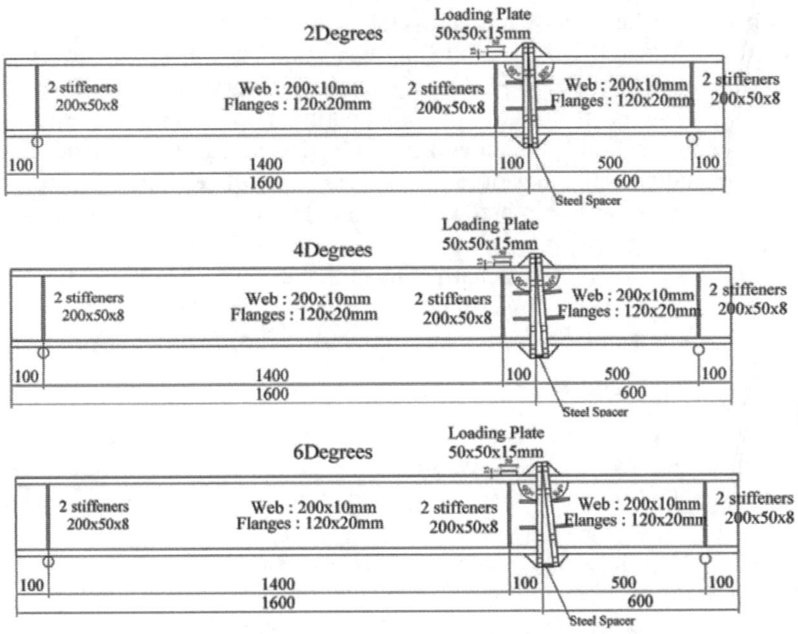

Figure 4. I240 BUS with initial imperfection tested specimens as presented by [11]

Stiffeners were provided at location of supports and at location of load application.

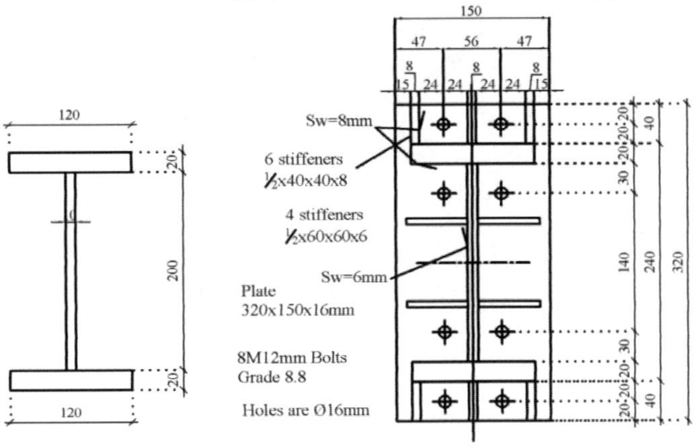

Figure 5. Beam cross section and specimens end plate details used in [11]

3.1. Finite Element Model Description

Two groups of models were developed to simulate the behavior of the connections using a finite element package, commercially known as ABAQUS v6.12 [12]. The first group of models, verification models, which were built to verify the FEM of the experimental program described by [11] using the load deflection behavior results discussed, while the other group was built for conducting the parametric studies, where multiple degrees of initial imperfection and different bolt diameters were considered for reaching comprehensive conclusions.

3.2. Material Properties Definition

For bolt material, the bolt tensile test data were available in [11], and the considered Poisson's ratio was ($\upsilon = 0.3$), the actual stress strain curve was obtained from the previously mentioned experimental program. While for steel sections properties, the yield stress of 360 MPa and ultimate stress of 520 MPa with ultimate strain at failure of 16% were considered.

3.3. Assigning Interaction Properties Between Parts

Tie constraint was considered when no slip or separation is allowed whatever the stress state or straining actions are, this was used between beam flanges and web, beams sections and end plates, and stiffeners and stiffened elements. While interactions were assigned to define the inter-surface behaviour, such as the interaction between end plates at the faying surfaces, and interaction between bolts and end plates including end plates outer surfaces and end plates holes circumferal area.
Generally in ABAQUS software, the interaction is defined by assigning behaviour in normal direction and in tangential direction, these behaviours were assigned as follows:

i. Normal behaviour: was assigned as hard contact that allows separation after contact; meaning it fully transfers compression stresses but doesn't transfer tensile stresses.
ii. Tangential behaviour: using coefficient of friction of 0.3 with sliding permitted.

3.4. Load Application

In order to precisely simulate the actual loading sequence, where bolts are first tightened before full utilisation of the structure, two steps of loading were defined, the first one is the step in which pretension force was applied on bolts, then the second step is the one in which the external load was gradually applied.

3.5. Element Type

The degrees of freedom are the fundamental variables calculated during the analysis, displacements or rotations are calculated at the nodes of the element. At any other point in the element, the displacements are obtained by interpolating using the nodal displacements. Hence, the higher number of nodes means more computational cost of the analysis.

ABAQUS software package has the capability of mixing different meshes types together in the same model, where low computational cost elements, like the tetrahedral elements, can be used in the model in regions of low interest, while regions of high interest can be modelled with higher computational cost elements.

In this paper the FE modeling of connections was developed using continuum (solid) elements for all parts of the connection. This kind of elements is capable of providing sufficient degrees of freedom to model contact, deformations and plasticity effects.

Because of their capabilities to mesh highly complex geometries with greater ease than C3D8 elements, the C3D6 wedge elements or tetrahedral C3D4 elements (Fig. (6)) were used to model beams, stiffeners, and end plates, while C3D8 elements were used for modeling the bolts.

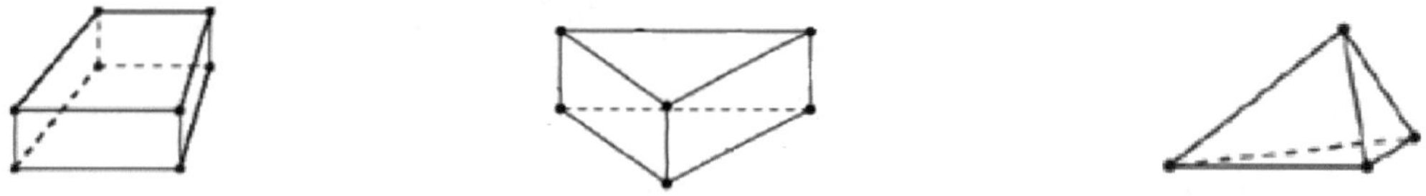

Figure 6. a) C3D8 solid elements. b) C3D6 solid elements c) C3D4 solid elements [12]

After some trials, the suitable mesh size for the bolts was found to be 3-4 mm, while for end plates region of contact the mesh size could be increased to be of 18-20 mm in order to reduce computational cost, and size of the mesh could be increased to around 50-60 mm for the remaining beams portions. Assembled model is shown in Fig. (7).

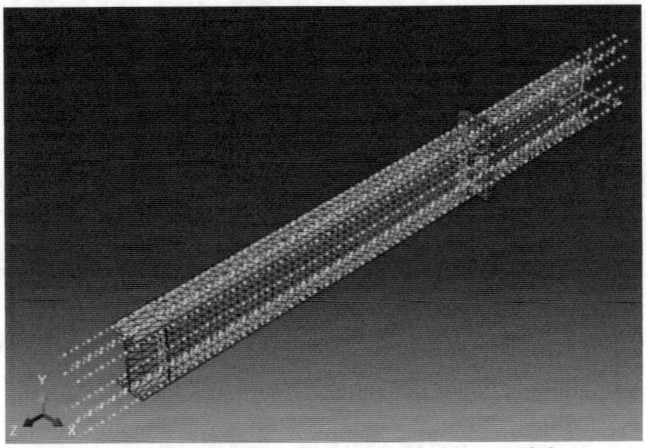

Figure 7. Assembled beam model

3.6. Results Comparison

3.6.1. Failure Load Values

The comparison between experimental failure load of the specimens tested by Abdallah et. Al [11] and failure loads acquired from FEM are presented in table (1), while model deformed shapes are shown in Fig. (8):

Table 1. I240 specimens experimental and FEM failure loads

	I240 Ideal	I240 2°	I240 4°	I240 6°
Experimental Failure Load (kN) Abdallah et. al [11]	151	129	126	118
Current FEM Failure Load (kN)	160.1	138.1	136.8	134.6
Percentage Difference (%)	6%	7.1%	8.5%	14%

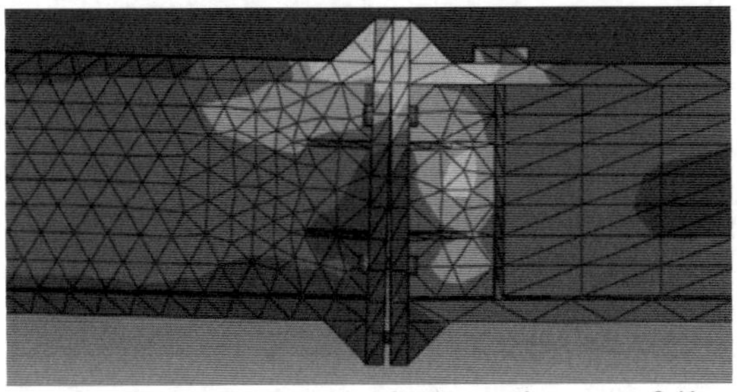

Fig. 8. a) FEM of I240 Ideal specimen at failure

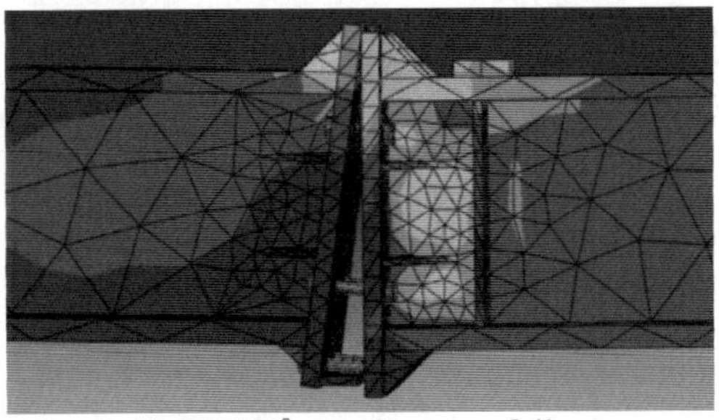

Fig. 8. b) I240 6° specimen at failure

3.6.2. Moment Deflection Curves

The comparison between experimental and FEM moment deflection curves are presented in Fig. (9-12).

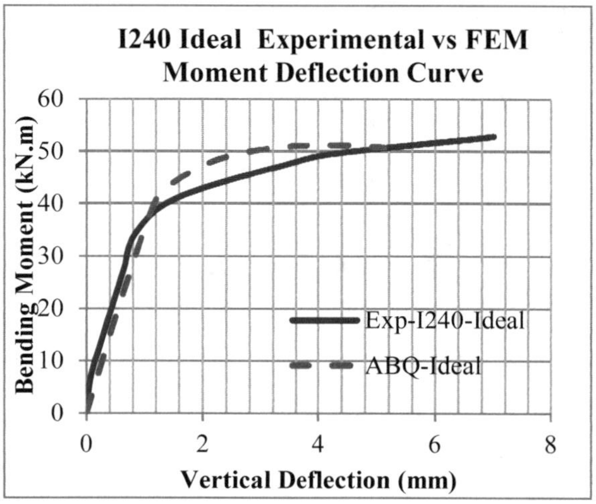

Figure 9. Exp. vs FEM of I240 Ideal specimen

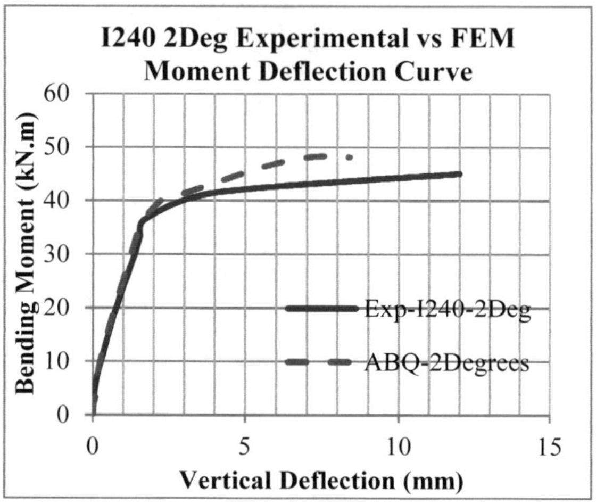

Figure 10. Exp. vs FEM of I240 2° specimen

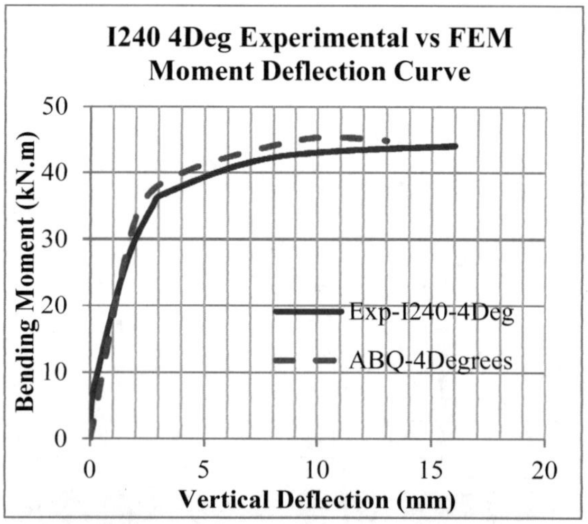

Figure 11. Exp. vs FEM of I240 4° specimen

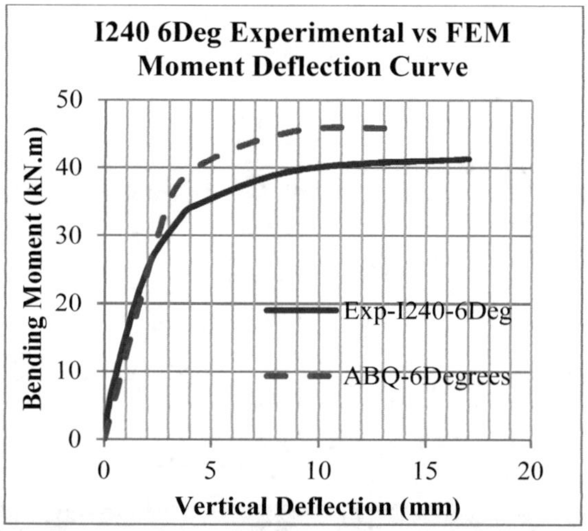

Figure 12. Exp. vs FEM of I240 6° specimen

3.7. Verification Process Conclusion

The comparison between FEM and experimental results revealed good agreement, which ensures the accuracy and suitability of the developed FEM for performing the numerical study on reliable basis.

4. Numerical Study of Imperfect Connections

4.1. Numerical Study Introduction

To further investigate the behavior of the initially imperfect connections, a group of connections with various parameters were modelled using FEM. The main purpose of this numerical study is to discuss the effect of the connections with initial imperfections gap angle on the behavior of the steel structure in terms of moment deflection relation.

4.2. Numerical Study Outlines

The structure considered was a cantilever beam connected to a column with the connection of interest, where the variables considered were bolt diameter and degree of initial imperfection, a layout of the connections considered is presented in Fig. (13), and the details of the connections are presented in Fig. (14).

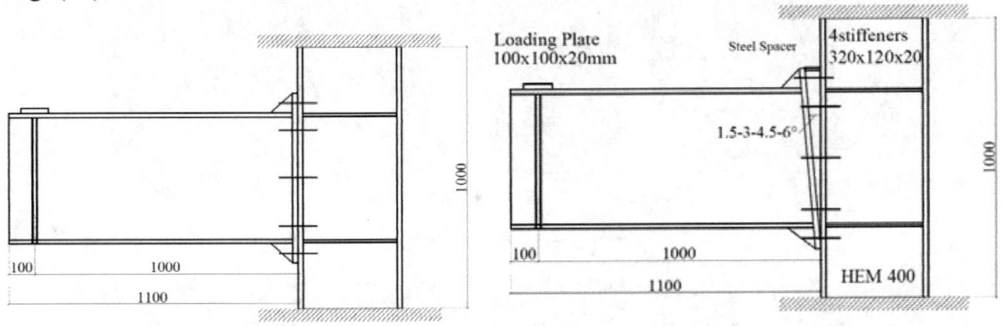

Figure 13. Layout of connections considered in parametric study

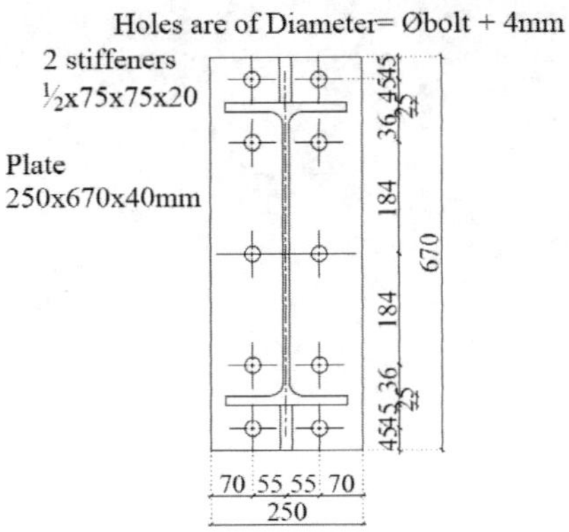

Figure 14. Details of connection considered

Beam used: a built-up section of flanges 200mm in width, 25mm in thickness, while web was 450mm in length and 16mm in thickness. Beam was of length 1100mm while loading was at distance 1000mm from the column external face, the beam was provided with sufficient stiffening below loading point to prevent excessive stresses. Column used was HEM 400 of length 1000mm, that was fixed at the start and at the end. The column was provided with sufficient stiffening at the connection region, right opposite to the beam flanges to prevent excessing stressing.

In order to effectively investigate the imperfect connection, four values of degrees of initial imperfection were considered; 1.5°, 3°, 4.5° and 6°, each with two values of bolts diameters were considered; 20&22mm, bolts were HSB type grade 8.8.

The summary of the parameters is presented in the following table.

Table 2. Numerical study summary and model nomenclature:

	Ideal	1.5°	3°	4.5°	6°
M20	M20-I	M20-1.5	M20-3	M20-4.5	M20-6
M22	M22-I	M22-1.5	M22-3	M22-4.5	M22-6

An illustration of the assemblies and meshed models are presented in Fig. (15-16).

IOP Conf. Series: Materials Science and Engineering 1150 (2021) 012020 doi:10.1088/1757-899X/1150/1/012020

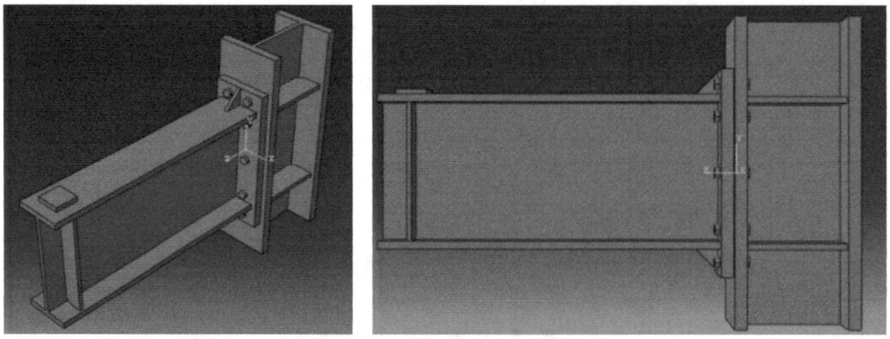

Figure 15. Assembled model

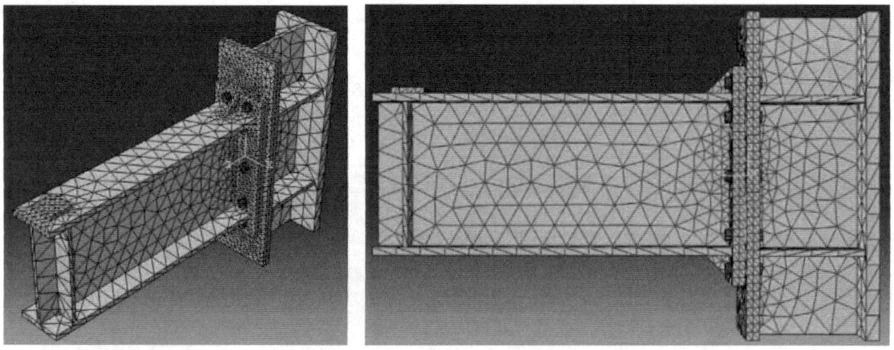

Figure 16. Meshed model

4.3. Numerical Study Results

A sample model state of stress is presented in Fig. (17-18), while a summary of the numerical study results is presented in table (3).

Figure 17. Connection M20-I stress state at maximum load (MPa)

IOP Conf. Series: Materials Science and Engineering 1150 (2021) 012020 doi:10.1088/1757-899X/1150/1/012020

Figure 18. Connection M22-6° stress state at maximum load (MPa)

Table 3. Numerical study results:

		Ideal	1.5°	3°	4.5°	6°
M20	Max. Deflection (mm)	14.8	18.7	25.5	29.6	31.1
	Ult.Moment (kN.m)	688.4	609	592.2	591.2	536.8
	% loss in capacity	—	11.53%	13.97%	14.12%	22.02%
M22	Max. Deflection (mm)	16	22.5	28	34.4	33.7
	Ult.Moment (kN.m)	794.8	685.1	667.1	650.2	641.2
	% loss in capacity	—	13.8%	16.06%	18.19%	19.32%

To better assess the effect of initial imperfections on connection, moment deflection curves of the imperfect connections as well as their ideal counterparts were plotted on the same graphical representation as shown in Fig. (19-20)

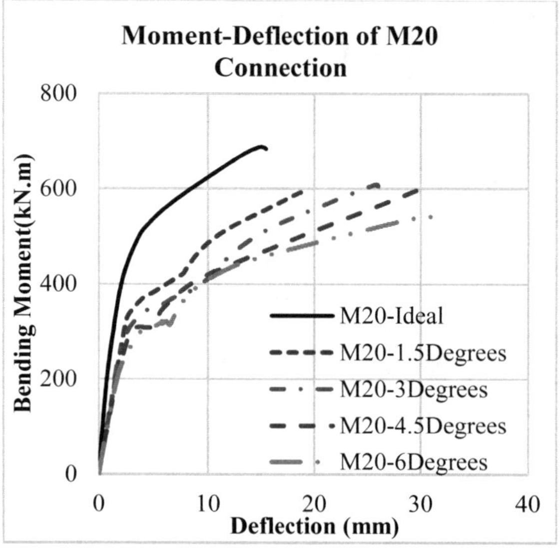

Figure 19. M20 group moment deflection curve

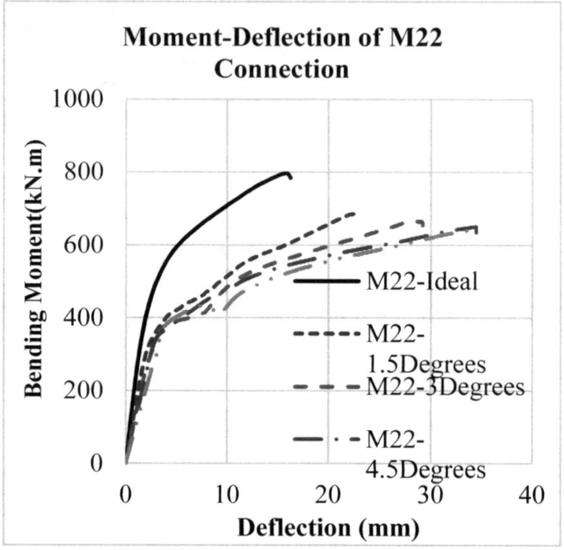

Figure 20. M22 group moment deflection curve

In real life structures, the occurrence of the imperfect connections is unplanned, one can obtain deflection and moment capacity of an ideal connection but not of an imperfect connection, so for better normalization, the actual deflection of the structure with initially imperfect connection was divided by the corresponding deflection of the ideal connection structure at same load level, then the acting load was divided by the ultimate moment capacity of the connection, and these two ratios ($\frac{M_{Acting}}{M_{Ultimate\ of\ Ideal}}$ (%) and $\frac{\Delta_{Imperfect}}{\Delta_{Ultimate\ of\ Ideal}}$ (%)) were plotted as shown in Fig. (21-22).

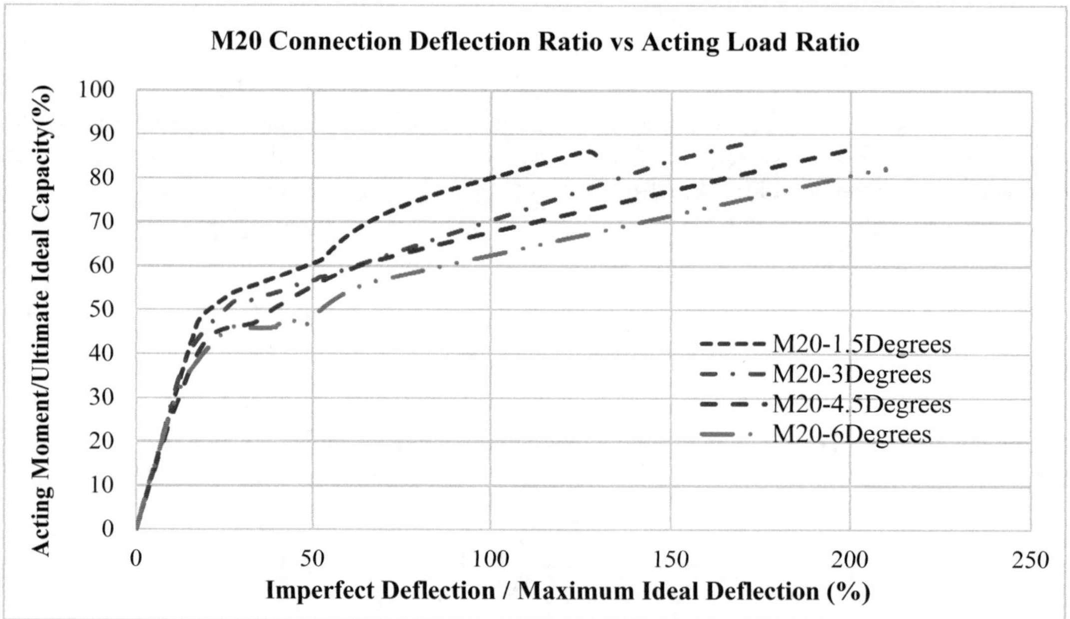

Figure 21. M20 connection group moment ratio vs deflection ratio

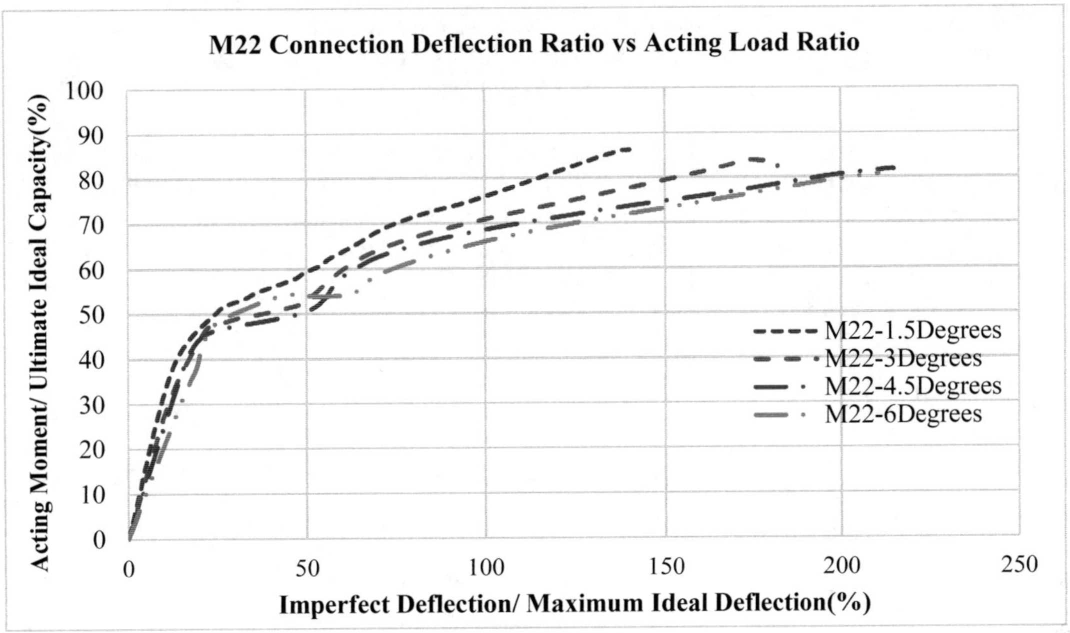

Figure 22. M22 connection group moment ratio vs deflection ratio

Conclusions

1. A reduction in load capacity of the initially imperfect connection ranging from 11.5% to 22.1% than that of the ideal connection with same bolts was noticed.
2. An increase in maximum deflection of the imperfect connections ranging from 26.35% to 110.6% more than ideal ones were noticed i.e.; more ductile behavior.
3. Maximum deflection increases with increasing the degree of initial imperfection.
4. The beam deflection of the imperfect connections is of larger magnitude compared to ideal counterpart at same moment level with a percentage ranging from 40.6% to 110.1%. i.e.; the stiffness of imperfect connection is less than the ideal ones.

Reference

[1] Reidar Bjorhovde, Andre Colson, and Jacques Brozzetti. " Classification system for beam to column connections" Journal of Structural Engineering, Vol. 116, No. 11, November, 1990. ©ASCE.
[2] Bahaari MR, Sherbourne AN. "Computer modelling of an extended endplate bolted connection". Computers and Structures 1994;52(5):879–93 .
[3] Baharri MR, Sherboure AN. "Finite element prediction offend plate bolted connection behavior II: Analytical formulation". Journal of Structural Engineering, ASCE 1997;123(2):165–75 .
[4] Masika RJ, Dunai L. "Behaviour of bolted end-plate portal frame joints". Journal of Construction Steel 1995;32(2):207–25.
[5] Wheeler AT, Clarke MJ, Hancock GJ. "FE modelling of four-bolt tubular moment end-plate connections". Journal of Structural Engineering, ASCE 2000;126(7): 816–22
[6] Shiming Chen, Gang Du (2007). "Influence of initial imperfection on the behaviour of extended bolted end-plate connections for portal frames" Journal of Constructional Steel Research 63 (2007) 211–220.
[7] H. J. Duana, J. C. Zhao and Z. S. Song "Effects of Initial Imperfection of Bolted End-plate Connections in the Reliability of Steel Portal Frames" Procedia Engineering 14 (2011) 2164–2171.

[8] Murray, T., M. and Shoemaker, W.L., "Steel Design Guide Series 16, Flush and Extended Multiple- Row Moment End-Plate connections", AISC, Chicago, IL, 2002.
[9] American Institute of Steel Construction (AISC), Steel Design Guide Series 4 (2003), "Extended End-Plate Moment Connection,", Chicago, IL.
[10] Eurocode 3: Design of steel structures- Part 1-8: Design of joints", ISBN 0580 46081 9.
[11] Mostafa Nour Eldin M., Emad Salem, Sherif M. Ibrahim, Abdelrahim K.Dessouki, Experimental Investigation on Initially Imperfect Rigid Connections, International Journal of Advanced Research in Engineering and Technology (IJARET) Volume 11, Issue 12, December 2020, pp.474-483, Article ID: IJARET_11_12_051 DOI: 10.34218/IJARET.11.12.2020.051
[12] ABAQUS Documentation Analysis User's Manual (2006).

Study of the physical-mechanical properties of igneous stones from Tlalpujahua Michoacan, Mexico

Elia Mercedes Alonso-Guzman[1,2, a *], Marco Antonio Navarrete-Seras[1,2,b], Wilfrido Martinez- Molina[1,2,c], Hugo Luis Chavez-Garcia[1,2,d], Cindy Lara-Gomez[1,b], José Carmen Garcia-Perdomo[1,e]

[1]Department of Materials, Faculty of Civil Engineering, Universidad Michoacana de San Nicolás de Hidalgo, Morelia, Michoacán, México, 58070
[2]Department of Basic Sciences and Mathematics, Faculty of Civil Engineering, Universidad Michoacana de San Nicolás de Hidalgo, Morelia, Michoacán, México, 58070

E-mail: elia.alonso@umich.mx

Abstract. Research work was carried out on rocks obtained in the municipality of Tlalpujahua Michoacán, located in different quarries of stone material, classifying the name of the quarry with the name of the community where the stones are located: "El moral" stones of bronze (BR) and golden (D) color, "la américa" stones of gray (G) and pink (R) color, "llano grande" stones of dark brown (CO), reddish-brown (CR) and black (N) color. Physical and mechanical tests were carried out, evaluating the percentage of moisture absorption, density, and absolute density. To evaluate the mechanical properties, the following tests were performed: uniaxial compressive strength (*UCS*), and static flexural test (modulus of rupture). The comparison of the different physical-mechanical tests allows estimating their behavior as a structural and restoration element. The correlation obtained to estimate the UCS has an acceptable correlation coefficient.

1. Introduction

Igneous stones are widely used for the conservation and restoration of historical monuments in Michoacan. Therefore, the study of the rock banks located in Tlalpujahua Michoacan, Michoacan, Mexico, is the subject of study in terms of their physical-mechanical properties. It is important to point out that nowadays this type of stone is also used in modern constructions such as houses, as a decorative element, or as a constructive element. Several researchers have studied different types of stones obtaining their physical and mechanical characteristics: Li et al. (2021) discuss the effects of grain size, grain morphology, and mineralogy on macro mechanical properties including crack initiation stress, crack damage stress, uniaxial compressive strength and elastic modulus in rocks [1]. Veríssimo-Anacleto et al. (2020) studied the uniaxial compressive behavior of limestone masonry with a different number of courses using small specimens. These results allowed the development of an empirical model to predict the compressive strength [2]. Navarrete-Seras et al. (2013,2020) studied the physical-mechanical behavior of quarries of materials from the Morelia Michoacán region [3-5]. Wang et al. (2018) carried out uniaxial compression tests (UCS) to investigate the influence of pre-existing crack geometry parameters (various dip angles, lengths, widths and numbers) on the

mechanical properties, and deformation failure modes of low-strength stone samples [6]. Niu et al. (2020) studied the classification of fracture modes of rocks during the cracking process by conducting uniaxial compression tests on intact and defective red sandstone samples [7]. Zhang et al. (2020) studied the deformation and fracture propagation of a sandstone sample under uniaxial compression using distributed fiber optic strain sensing (DFOSS) technology [8]. Ludovico-Marques et al. (2012) carried out an experimental investigation, based on uniaxial compression tests, on selected representative samples of sandstone and granite rock lithotypes used in the historic built heritage in Portugal, to evaluate the mechanical behavior under compression of different building stones [9].
In this work, different stones from Tlalpujahua Michoacán Mexico were studied, obtaining uniaxial compressive strength, static bending in stones, density, and absorption, these results were compared and acceptable correlation coefficients were obtained, the study is important due to the use of this stone in the remodeling and restoration of historical monuments in the region.

2. Methodology

The physical characterization consists of different tests to determine the content of the test analysis; the percentage of moisture absorption and absolute density. Color samples were obtained from three quarry stones, collecting seven colors, five cubic samples of 5cm x 5cm x 5cm were analyzed, giving a total of 35 specimens.

2.1. *Percentage of moisture absorption*

The absorption moisture content test was carried out according to the standards indicated in ASTM C127-04 and the Materials Analysis Book of the "Ing. Luis Silva Ruelas" Laboratory since they coincide in the objective and procedure [10,11].
The percentage of water that can be absorbed by a representative sample of rock during a time of 24 hours was determined concerning its dry mass.

2.2. *Absolute density through the pycnometer*

The objective is to determine the density of a stone of known mass by volume displacement. On the other hand, the mechanical tests that were performed are defined as the material's capacity to resist external or internal actions that involve the application of forces on it. They are the most common ones found among this type of tests and are very important for the characterization of the material, these forces are; Uniaxial Compressive Strength (*UCS*) and static bending in rocks (*MR*) [12].

2.3. *Uniaxial compressive strength (UCS)*

To determine what is the maximum load over the area that a rock can support [11,13]. The test was performed with the Tinius Olsen or Forney Universal Testing Machine.

2.4. *Static flexural strength in stones*

Determine the modulus of rupture and shear stress of rocks with standardized dimensions to be met by the specimens to be tested and the necessary accessories to determine the flexural strength (figure 1) [11].

The flexural modulus of rupture of a stone equation (1) is determined.

$$MR = \frac{3WL}{2BD^2}\,, \tag{1}$$

and the shear stress is calculated equation (2).

$$\sigma = \frac{3W}{2A} = \frac{3W}{2BD}\,, \tag{2}$$

σ is shear stress (Kg/cm^2), W is the maximum load applied to the specimen (kg), L is the length between supports (cm), A is specimen section area (cm^2), B is specimen width (cm) and D is specimen camber (cm).

Figure 1. Static bending test performed on the hydraulic dam (Source: J.C. Garcia Perdomo).

3. Results and discussion

Table 1 shows the percentage of moisture absorption obtained from the different rock colors. The bronze (BR) and pink (R) colors have low absorption, which is proportional to the density, that is, they have low porosity content.

While the golden (D) and gray (G) colors show higher absorption, i.e. low density and higher amount of pores.

As for the density, it can be observed that the densities are in the range of 2.1 and 2.24 (g/cm^3) (figure 2). According to the measured density of the material, indicating the SCT standard designation M-MMP-2-01-005/02, the apparent density of masonry materials is at least 2.3 g/cm^3, and the results obtained comply with the density to be used for masonry. However, it was observed that the bronze-colored stone (BR) presented better results.

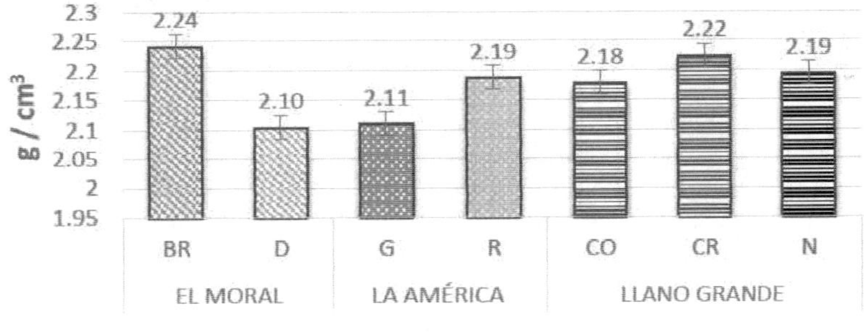

Figure 2. Absolute density graph.

Table 1 shows the results of uniaxial compressive strength in quarry cubes of different colors, with the highest strength response among the test specimens in the study being pink and the lowest strength response being reddish brown. The "La América" bench has material with higher resistance to uniaxial compression. According to the SCT standard designation M-MMP-2-01-002/02, in compressive strength of masonry materials, when it contains sedimentation, shale, or fracture planes due to tectonism, the minimum strength in the direction perpendicular to these planes will be 153 Kg/cm^2, and 101 Kg/cm^2, in the parallel direction, which indicates that all stones comply with the effort subjected to compression in masonry [14]. According to the International Society of Rock Mechanics (ISRM, 1981), it is classified by bronze-colored stone (BR) from "el moral" quarry, gray (G) from "la América" quarry and the dark brown (CO), reddish-brown (CR), and black (N) stones from "Llano grande" bank as soft rock, where they are within a parameter of 5 and 25 MPa and the golden (D) and pink (R) colored stones as moderately hard stone, they are within a parameter of 25 and 50 MPa.

As for the modulus of rupture in the prismatic specimens, the gray-colored specimens have the highest strength and the pink-colored specimens have the lowest flexural strength.

Table 1. Results of physical and mechanical properties, an average of five samples.

ORIGIN OF THE STONE	COLOR	% ABSORPTION	DENSITY (g/cm^3)	UCS (MPa)	MR (MPa)
El Moral	BR	12.35	2.24	20.26	7.46
	D	16.44	2.1	27.09	7.72
La América	G	17.6	2.11	23.01	9.38
	R	13.28	2.19	32.19	6.09
Llano grande	CO	15.68	2.18	14.62	7.13
	CR	15.68	2.22	14.2	8.41
	N	15.64	2.19	19.13	7.01

On the other hand, the coefficient of determination can be used to evaluate the goodness of fit of the model and always varies between 0 and 1 [15]. If R^2 is equal to 0.756, it means that approximately 75.6% of the variation in Y is due only to the linear regression of Y on X [16].

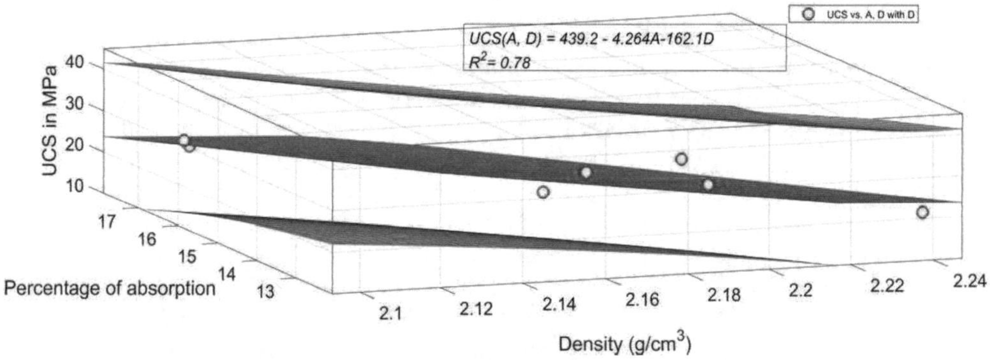

Figure 3. Linear regression model to estimate the uniaxial compressive strength (*UCS*) for the stones under study.

In figure 3 a linear regression model can be observed, the data are fitted to a surface known as hyperplane, in this model the coefficient of determination R^2 is 0.78, also, the 95 percent confidence limits can be observed.

4. Conclusions

According to the results obtained, the absolute density of the quarry material varies between 2.10 and 2.24 g/cm^3, being slightly below the minimum specification of 2.3 g/cm^3 for the use of masonry materials.

In the uniaxial compressive strength (*UCS*) in cubes, the best results were obtained in pink-colored rocks and lower in reddish colored ones, however, all the evaluated samples are considered efficient for structural use for masonry and among other elements to be used.

The "Llano Grande" bench shows its best results in bending stress in black material, being also the one that is mostly used for moldings, sculptures, and laminates. The samples that were less favorable in the evaluation for static and shear bending in stones can be improved by increasing the thickness of the stone.

The linear mathematical model by fitting the data to the hyperplane allows estimating the uniaxial compressive strength since it has an acceptable correlation coefficient.

5. References

[1] X.F. Li, H.B. Li, J. Zhao, Transgranular fracturing of crystalline rocks and its influence on rock strengths: Insights from a grain-scale continuum–discontinuum approach, Comput. Methods Appl. Mech. Eng. 373 (2021) 113462. https://doi.org/10.1016/j.cma.2020.113462.

[2] J. Veríssimo-Anacleto, M. Ludovico-Marques, P. Neto, An empirical model for compressive strength of the limestone masonry based on number of courses – An experimental study, Constr. Build. Mater. 258 (2020) 119508. https://doi.org/10.1016/j.conbuildmat.2020.119508.

[3] M.A. Navarrete Seras, F.J. Domínguez Mota, E.M. Alonso-Guzmán, W. Martínez-Molina, H.L. Chavez-Garcia, J.G. Tinoco Ruiz, J.A. Guzmán Torres, J.E. Arreygue Rocha, Slake Durability Index of the Stone Aggregate Used in the Morelia Michoacán Region, Key Eng. Mater. 862 (2020) 61–65. https://doi.org/10.4028/www.scientific.net/KEM.862.61.

[4] M.A.N. Seras, F.J.D. Mota, E.M.A. Guzmán, W.M. Molina, H.L.C. García, J.G.T. Ruiz, J.A.G. Torres, Scanning Electron Microscope in Rocks and their Comparison with Physical-Mechanical Properties, Key Eng. Mater. 841 (2020) 114–118. https://doi.org/10.4028/www.scientific.net/KEM.841.114.

[5] M. Navarrete, W. Martínez Molina, E.M. Alonso-Guzmán, C. Lara-Gómez, J.A. Bedolla-Arroyo, H. Chávez, D. Delgado, J.C. Arteaga, Caracterización de propiedades físico-mecánicas de rocas ígneas utilizadas en obras de infraestructura, Rev. ALCONPAT. 3 (2013) 129–139. https://doi.org/10.21041/ra.v3i2.49.

[6] Y. Wang, J. Tang, Z. Dai, T. Yi, Experimental study on mechanical properties and failure modes of low-strength rock samples containing different fissures under uniaxial compression, Eng. Fract. Mech. 197 (2018) 1–20. https://doi.org/10.1016/j.engfracmech.2018.04.044.

[7] Y. Niu, X.-P. Zhou, F. Berto, Evaluation of fracture mode classification in flawed red sandstone under uniaxial compression, Theor. Appl. Fract. Mech. 107 (2020) 102528. https://doi.org/10.1016/j.tafmec.2020.102528.

[8] L. Zhang, D. Yang, Z. Chen, A. Liu, Deformation and failure characteristics of sandstone under uniaxial compression using distributed fiber optic strain sensing, J. Rock Mech. Geotech. Eng. 12 (2020) 1046–1055. https://doi.org/10.1016/j.jrmge.2019.12.015.

[9] M. Ludovico-Marques, C. Chastre, G. Vasconcelos, Modelling the compressive mechanical behaviour of granite and sandstone historical building stones, Constr. Build. Mater. 28 (2012) 372–381. https://doi.org/10.1016/j.conbuildmat.2011.08.083.

[10] A.S. D18.12, Standard Test Method for Laboratory Method for Density, Relative Density (Specific Gravity), and Absorption of Coarse Aggregate ASTM C127-04, American Society for Testing Materials (ASTM), 2004.

[11] L. Navarro, W. Martínez, A. Espinoza, ANÁLISIS DE MATERIALES, Manual de Análisis de Materiales, Resistencia de Materiales, 2011.

[12] L.-I. González-De-Vallejo, M. Ferrer, Geological Engineering, CRC Press/Balkema, 2011.

[13] A.S. D18.12, Standard Test Method for Unconfined Compressive Strength of Intact Rock Core Specimens (ASTM D2938-95(2002)), American Society for Testing Materials (ASTM), 2002.

[14] N-CMT-2-01-003, Materiales para mamposterías. SCT, n.d.

[15] A. Silva, J. de Brito, P.L. Gaspar, Methodologies for Service Life Prediction of Buildings With a Focus on Façade Claddings, Springer, Cham, 2016. https://doi.org/https://doi.org/10.1007/978-3-319-33290-1.

[16] C.S. Rayat, Statistical Methods in Medical Research, Springer Singapore, Singapore, 2018. https://doi.org/10.1007/978-981-13-0827-7.

An 18 slot, 8-pole Permanent Magnet Synchronous Generator for Small-Scale Wind Turbine Energy Conversion: Design, Fabrication, and Experiments

Natalina Damanik[1,a*], Sahrijal Purba[2,a], M. Al Roshady S. Calle[3,b], M. Arie Kurniawan[4,b]

[a]PT PLN (Persero) PUSLITBANG Ketenagalistrikan (Research Institute), The Indonesia State Electricity Company, 12760 Jakarta, Indonesia
[b]Lentera Bumi Nusantara, Research Centre for Renewable Energy, Ciheras, Kabupaten Tasikmalaya, West Java, Indonesia

E-mail: natalina@pln.co.id

Abstract. Compact permanent magnet synchronous generators (PMSGs) for renewable energy conversion are required for remote areas. In this work, an 18-slot, 8-pole PMSG (18S8P PMSG) was designed and fabricated for small-scale wind turbine energy conversion. Finite element analysis was conducted to assess the magnetic flux density distribution of the 18S8P PMSG prototype. The simulation results showed that most of the magnetic flux values were within 0.4–1.5 T, which was well below the saturation level for 35JN360 material. The magnetic flux intensities of the stator and rotor cores were within the optimal magnetic permeability range of the 35JN360 material. There were seven steps involved in fabricating the 18S8P PMSG prototype: (1) material cutting, (2) burring, (3) stacking, (4) welding, (5) coil winding, (6) configuration wiring, and (7) assembling. The electromagnetic performance results showed that the back electromotive force voltage of the 18S8P PMSG prototype seemed to be proportional to the rotor speed, where K_E was 0.52 V·s/rad. The no load torque maximum of 18S8P PMSG is 1.7 N.m at 5000 rpm. The mechanical performance results showed that the highest power output (2500 W) was obtained for the 18S8P PMSG prototype at a rotor speed of 5000 rpm using the 20-Ω load model. The maximum efficiency of this prototype was 85.4%.

1. Introduction

The limited fossil fuel reserves and the growing awareness for clean energy resources has spurred the development of renewable energy technologies over the years. Small-scale wind turbines are the best solution for remote areas (which are located far from the power grid) as well as archipelagos such as Indonesia. In wind turbines, the wind power is converted into mechanical power through the aerodynamic effect of the rotor blades. Many studies have been carried out to optimize the structure and performance of permanent magnet synchronous generators (PMSGs).

Verde, Lastres [2018] Yousefian and Meshgin Kelk [2018]. To design a good PMSG, the following criteria need to be fulfilled: (1) the power output, torque, and efficiency need to be maximized and (2) the cogging torque needs to be minimized [3]. In this work, an 18-slot, 8-pole PMSG (18S8P PMSG) prototype was designed and fabricated. Two-dimensional FEA was conducted to assess the magnetic flux distribution of the 18S8P PMSG internal structure. Finally, experiments were carried out to evaluate the electromagnetic and mechanical performance of the 18S8P PMSG prototype.

2. Methodology

The methodology adopted in this work consisted of the following phases: (1) design of the 18S8P PMSG prototype, (2) design, (2) FEA (which is a numerical method used to solve linear and nonlinear partial differential equations [4]) to assess the magnetic flux density distribution of the 18S8P PMSG internal structure, (3) fabrication, and (4) experimental testing to evaluate the electromagnetic and mechanical performance. The 18S8P PMSG prototype was designed based on the specifications listed in Table 1. The schematic of the 18S8P PMSG internal structure is shown in Fig. 1(a). The two-dimensional finite element model of the 18S8P PMSG internal structure was meshed, as shown in Fig. 1(b). The 18S8P PMSG assembly model is shown in Fig. 2. The 18S8P PMSG prototype was fabricated according to the following steps. First, the core material was cut using a laser cutting machine. Second, the housing was fabricated using a computer numerical controlled (CNC) machine. Third, the rotor shaft was formed using a lathe machine. Lastly, the 18S8P PMSG prototype was assembled manually. The performance of the 18S8P PMSG prototype (i.e., electromagnetic and mechanical performance) was tested using the experimental setup shown in Fig. 3. The instruments and equipment used for the performance testing are presented in Table 2. The electromagnetic performance was assessed in terms of the back electromotive force (BEMF) constant and torque constant at a rotor speed of 500–5000 rpm. The BEMF constants was calculated using Eq. (1) [5]. The mechanical performance was assessed in terms of the power output and efficiency.

$$K_E = e/\omega_m \quad (1)$$

Here, K_E is the BEMF constant, e is the BEMF, ω_m is the rotor speed

Table 1. Design specifications of the 18S8P PMSG prototype.

Slots/Poles		18/8
Stator: outer diameter/inner diameter	(mm)	150/102
Rotor: Outer diameter	(mm)	100
.Air gap	(mm)	1
Stack length	(mm)	35
Coil specifications	(diameter/number of turns)	1.0mm/36
Coil connection		Y-3Ser2Par
Core material		35JN360
Magnet material		Neodymium NMX-48SH

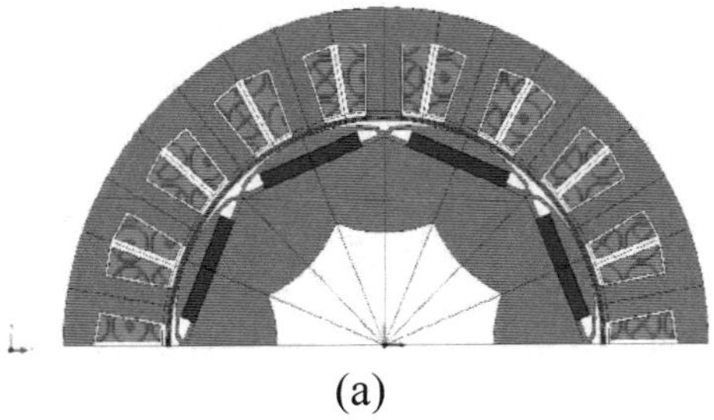

(a)

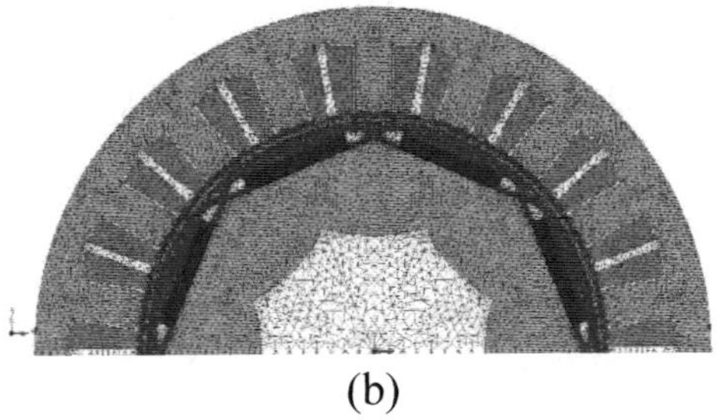

(b)

Figure. 1 (a) Schematic of the 18S8P PMSG internal structure. (b) Meshing of 18S8P PMSG model in FEA.

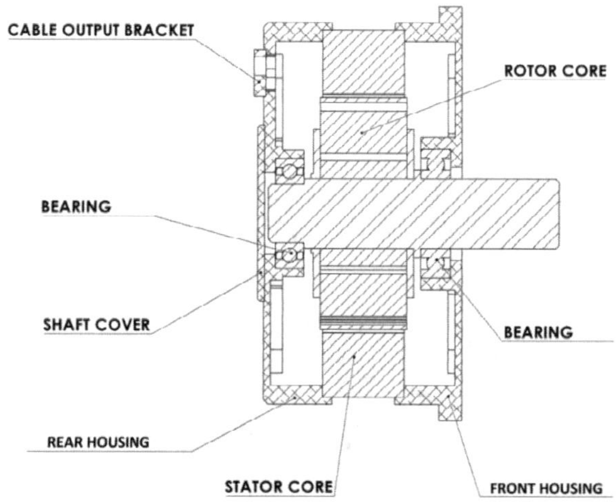

Figure. 2 Schematic of the 18S8P PMSG assembly model.

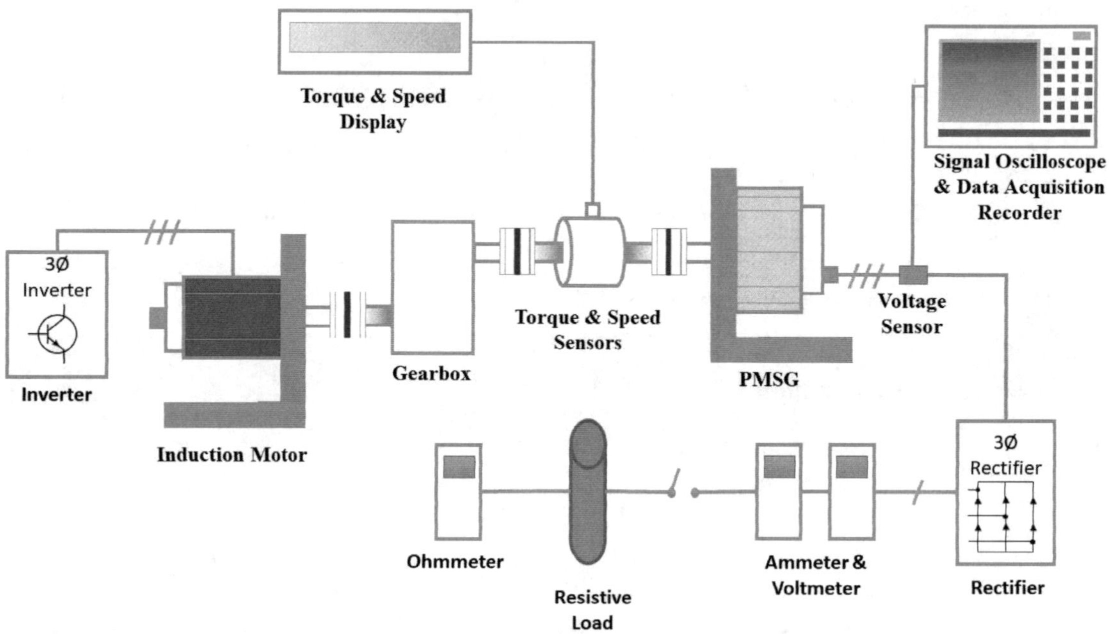

Figure. 3 Schematic of experimental setup used to test the performance of the 18S8P PMSG prototype.

Table 2. List of instruments and equipment used to test the performance of the 18S8P PMSG prototype.

No.	Measurement output	Instrument/Equipment	Brand
1	Voltage spectrum	Signal oscilloscope and data acquisition recorder	Scope corder DL850EV, Yokogawa Test and measurement Corporation
2	Torque and speed	Torque sensor and speed sensor	Magtrol SA Type:TM 310/011
3	Current	Ammeter	Fluxe
4	Voltage	Voltmeter	Fluxe
5	Resistive load	Ohmmeter	Fluxe

3. Results and discussion

3.1. Finite element analysis

Fig. 4 shows the magnetic flux density distribution of the 18S8P PMSG prototype, where the red and white colors indicate the maximal and minimal magnetic flux densities, respectively. In general, the magnetic flux density was within a range of 0–2 T, and the maximal magnetic flux density was observed around the rotor tip. However, most of the magnetic flux densities of the 18S8P PMSG were within a range of 0.4–1.5 T. It is essential to analyze the magnetic flux density distribution to ensure that the prototype design was within the design capability of the materials.

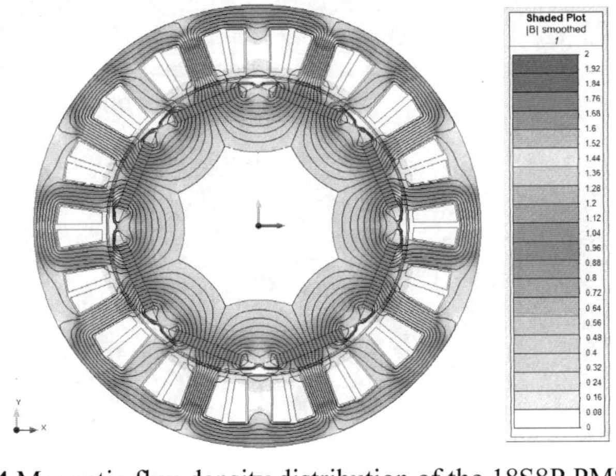

Figure. 4 Magnetic flux density distribution of the 18S8P PMSG prototype.

Fig. 5 shows the comparison between the magnetic flux intensity values of the 18S8P PMSG stator and rotor cores and the magnetic flux density curve of the 35JN360 material. The magnetic flux intensities of the 18S8P PMSG stator and rotor cores were within the optimal flux density range (0.1–1.5 T) of the 35JN360 material. The magnetic flux intensities of the 18S8P PMSG stator and rotor cores were well below the magnetic saturation level of the 35JN360 material. The saturation level of the ferromagnetic material resulted in a lower magnetic flux density even when the rotor speed of the 18S8P PMSG prototype was increased. According to Clegg, Beckley [2003] materials with lower saturation flux density requires a larger core cross-sectional area to carry the magnetic flux. Fig. 6 shows the comparison between the magnetic flux intensities of the 18S8P PMSG stator and rotor cores and the relative permeability curve of the 35JN360 material. Magnetic permeability is a measure of the ease with which magnetic flux flows through a material [7]. The magnetic flux intensities of the 18S8P PMSG stator and rotor cores were within the optimal relative permeability range of the 35JN360 material.

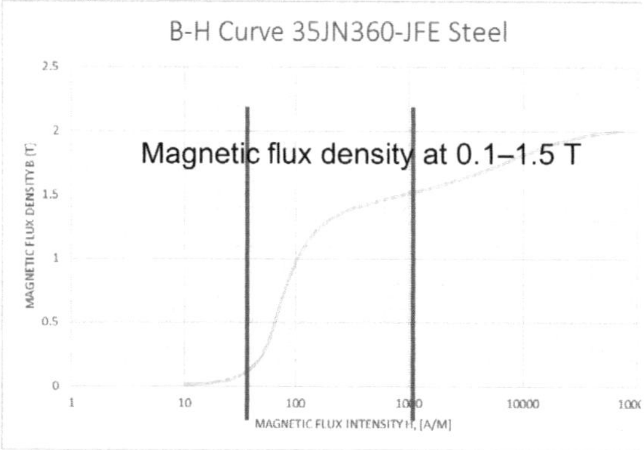

Figure. 5 Comparison between the magnetic flux intensity values of the 18S8P PMSG stator and rotor cores and the magnetic flux density curve of the 35JN360 material.

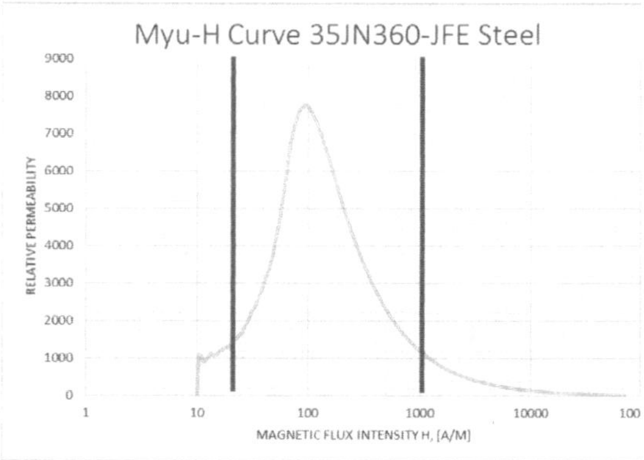

Figure. 6 Comparison between the magnetic flux intensity values of the 18S8P PMSG stator and rotor cores and the magnetic permeability of the 35JN360 material.

Fabrication of the 18S8P PMSG prototype

Fig. 7 shows the fabrication process of the 18S8P PMSG prototype. There were seven steps involved in fabricating the 18S8P PMSG prototype: (1) material cutting, (2) burring, (3) stacking, (4) welding, (5) coil winding, (6) configuration wiring, and (7) assembling. First, the 35JN360 material sheets (used for the stator and rotor cores) were cut using a laser cutting machine. The material sheets for the rotor and stator were stacked and welded so that the 18S8P PMSG components can be tightened at the designated points.

Figure. 7 Steps involved in fabricating the 18S8P PMSG prototype.

Analysis of the 18S8P PMSG prototype performance

Fig. 8 and 9 show the BEMF voltage and torque of the 18S8P PMSG prototype, respectively. Based on the results, both the BEMF voltage and torque appeared to be proportional to the rotor speed. BEMF is generated from changes in the magnetic flux and it can be expressed by the product of the BEMF constant and rotor speed. BEMF is the difference between the terminal voltage and the neutral voltage

IOP Conf. Series: Materials Science and Engineering 1150 (2021) 012022 doi:10.1088/1757-899X/1150/1/012022

of the non-energized phase because there is no current flow in the non-energized phase [5]. The BEMF constant, K_E was determined to be 0.52 V·s/rad. The no load torque maximum of 18S8P PMSG is 1.7 N.m at 5000 rpm. This torque is resulted by mechanical losses, cogging and core losses.

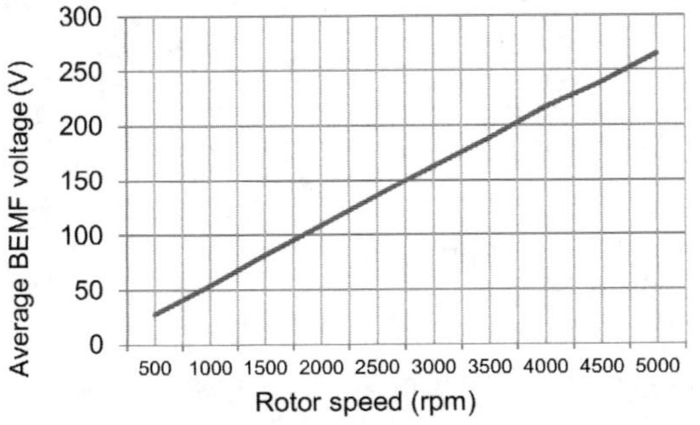

Figure. 8 BEMF voltage of the 18S8P PMSG prototype at different rotor speeds (500–5000 rpm).

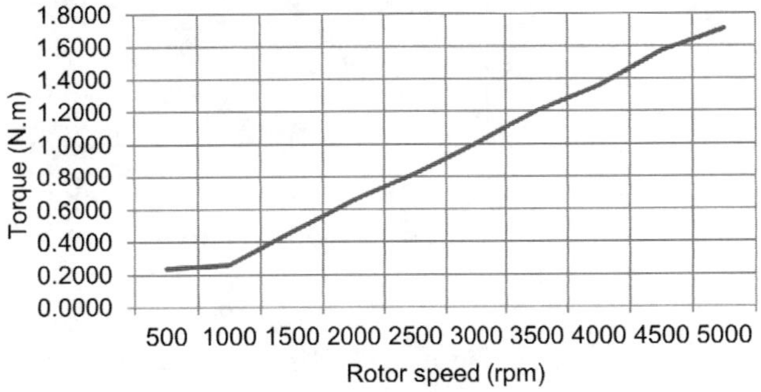

Figure. 9 Torque of the 18S8P PMSG prototype at different rotor speeds (500–5000 rpm).

Fig. 10 shows the power output of the 18S8P PMSG prototype at different rotor speeds. In general, the power output of the 18S8P PMSG increased as the rotor speed increased. For the lowest resistive load model (10 Ω), the highest power output was attained at 3500 rpm, with a value of 2000 W. The maximum power output (2500 W) was achieved for the 20-Ω load resistive model at a rotor speed of 5000 rpm. Owing to the wire current capacity limitation, the lowest resistive load model (10 Ω) was not tested at rotor speeds of more than 3500 rpm.

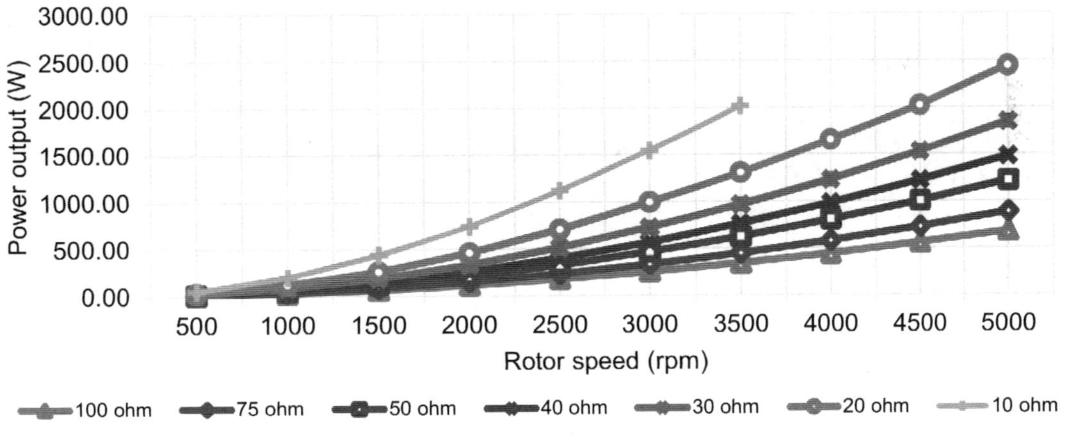

Figure. 10 Power output of the 18S8P PMSG prototype at different rotor speeds.

Fig. 11 shows the efficiency of the 18S8P PMSG prototype. Overall, the efficiency increased with an increase in the rotor speed. The efficiency of the 18S8P PMSG prototype was highest for the 20-Ω resistive load model whereas the lowest efficiency was obtained for the 100-Ω resistive load model. The highest efficiency of the 18S8P PMSG prototype was 85.4%.

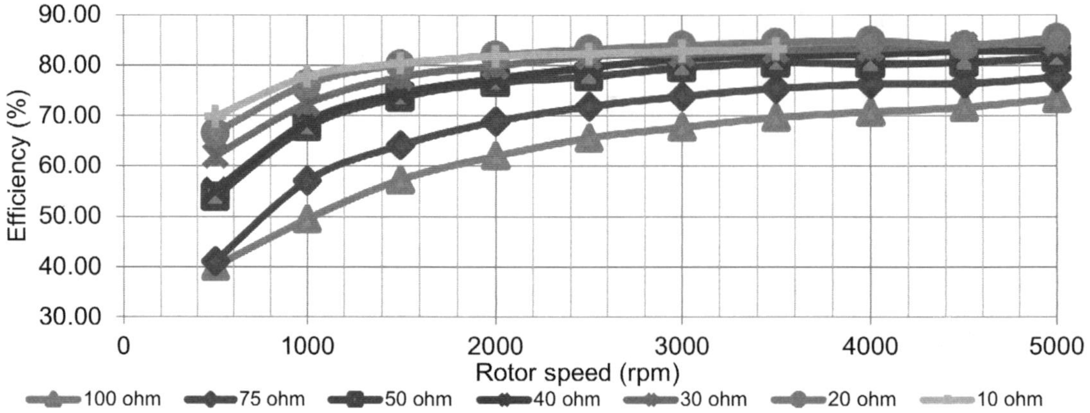

Figure. 11 Efficiency of the 18S8P PMSG prototype at different rotor speeds.

4. Conclusion

In this work, an 18S8P PMSG prototype was designed and fabricated for small-scale wind turbine energy conversion, and experiments were conducted to analyze its performance. Based on the FEA results, the magnetic flux density of the 18S8P PMSG prototype was within 0.4–1.5 T, which was well below the magnetic saturation level for 35JN360 material. The magnetic flux intensities of the 18S8P PMSG stator and rotor cores were within the optimal magnetic permeability of the 35JN360 material. There were seven steps involved in fabricating the 18S8P PMSG prototype: (1) material cutting, (2) burring, (3) stacking, (4) welding, (5) coil winding, (6) configuration wiring, and (7) assembling. The electromagnetic performance results revealed that the BEMF voltage and torque of the 18S8P PMSG prototype seemed proportional to the rotor speed, where K_E was found to be 0.52 V·s/rad. The mechanical performance results showed that the power output of the 18S8P PMSG prototype was highest (2500 W) at a rotor speed of 5000 rpm for the 20-Ω resistive load model. The maximum efficiency of the 18S8P PMSG prototype was 85.4%.

Acknowledgment

The authors acknowledge the PT PLN (Persero) Puslitbang Ketenagalistikan (Research Institute), South Jakarta, DKI Jakarta, Indonesia, for the financial support (KPP:20/LIT/BKIT/2018).

References

[1] Verde, A., et al., *A new method for characterization of small capacity wind turbines with permanent magnet synchronous generator: An experimental study.* Heliyon, 2018. **4**(8): p. 732.

[2] Yousefian, H.a. and H. Meshgin Kelk, *A unique optimized double-stator permanent-magnet synchronous generator in high-power wind plants.* Energy, 2018. **143**: p. 973-979.

[3] Jamali Arand, S. and M. Ardebili, *Cogging torque reduction in axial-flux permanent magnet wind generators with yokeless and segmented armature by radially segmented and peripherally shifted magnet pieces.* Renewable Energy, 2016. **99**: p. 95-106.

[4] Minaz, M.R. and M. Çelebi, *Design and analysis of a new axial flux coreless PMSG with three rotors and double stators.* Results in Physics, 2017. **7**: p. 183-188.

[5] Jang, G.H., J.H. Yeom, and M.G. Kim, *Determination of torque–speed–current characteristics of a brushless DC motor by utilizing back-EMF of non-energized phase.* Journal of Magnetism and Magnetic Materials, 2007. **310**(2, Part 3): p. 2790-2792.

[6] Clegg, A.G., et al., *8 - Magnetic Materials*, in *Electrical Engineer's Reference Book (Sixteenth Edition)*, M.A. Laughton and D.J. Warne, Editors. 2003, Newnes: Oxford. p. 8-1-8-17.

[7] Khazdozian, H.A., R.L. Hadimani, and D.C. Jiles. *Size reduction of permanent magnet generators for wind turbines with higher energy density permanent magnets*. in *2014 North American Power Symposium (NAPS)*. 2014.

IOP Conf. Series: Materials Science and Engineering 1150 (2021) 012023 doi:10.1088/1757-899X/1150/1/012023

Verification of Finite Element Computational Model for Biaxial Buckling of Stiffened Plates

Guilherme Ribeiro Baumgardt[1,a*], Thiago da Silveira [1,b], João Paulo Lima[2,c], Luiz Alberto Oliveira Rocha[3,d], Elizaldo Domingues dos Santos[1,e], and Liércio André Isoldi[1,f]

[1]Graduate Program in Computational Modeling (PPGMC), Federal University of Rio Grande (FURG), km 8 Itália Ave., Rio Grande, RS, Brazil
[2]Federal University of Goiás (UFG), 920 Mucuri St., Aparecida de Goiânia, Go, Brazil
[3]University of Vale do Rio dos Sinos (Unisinos), 950 Unisinos Ave., São Leopoldo, RS, Brazil

E-mail: guilhermebaumgardt@gmail.com

Abstract. This study presents the verification of computational models for the analysis of biaxial buckling in plates. The verification was performed by comparing the numerical results obtained in this study with results found in scientific references published by other authors. The computational models were developed in the ANSYS® software, which is based on the Finite Element Method (FEM), and the SHELL281 element is used. The mesh convergence tests demonstrate the convergent behavior of the simulations performed for the analyzed examples, which are: a square plate under biaxial elastic buckling, and a rectangular plate with stiffeners under biaxial elasto-plastic buckling. From the converged mesh, maximum differences of 0.84% and 3.89% were achieved when compared with references, respectively, for the elastic and elasto-plastic biaxial buckling. These results indicate that the proposed computational models were adequately verified.

1. Introduction

Thin plates are structural components widely used in the aircraft industry, the shipbuilding industry and civil construction. The relatively low compressive load that a thin plate supports can be increased through the use of stiffeners. The stiffeners are beams fixed transversely and/or longitudinally in the plane of the plate, increasing its structural rigidity [1]. This increase in stiffness can be proven through the results presented in [2] where, keeping the volume of material constant, significant reductions for the central deflection of a rectangular steel plate subjected to bending were achieved due the use of stiffeners

However, the inclusion of stiffeners in the plate generates geometric discontinuities, which makes its analysis complex or even impossible through analytical methods [3]; then the numerical methods are the most used approach for the analysis of its mechanical behavior.

In addition, it is well known that thin plates when subjected to compressive forces in the plane, can suffer buckling. But, unlike columns, the critical buckling load does not configure its collapse, since they have a stable post-buckling behavior [4].

In this sense, the behavior of plate buckling has been numerically investigated by several authors. Xu and Soares [5,6] proposed an experimental and numerical study via Finite Element Method (FEM), of uniaxial plate buckling with stiffeners, to assess the ultimate stress in the intermediate gaps of the plate, applying different configurations and boundary conditions. Through the comparison between the two studies, it was observed that the stress distributions in the reinforced panels obtained by the numerical model reached slightly higher magnitudes than the experimental tests, probably due to the residual stresses of the welding and details of the plate's manufacture. Lima et al. [7] performed the numerical study of the mechanical behavior of rectangular stiffened steel plates subjected to elastic and elasto-plastic buckling, under uniaxial compressive loading, using the Constructal Design Method associated with the Exhaustive Search technique and the FEM (by ANSYS® software) in a study of geometric optimization, evaluating the influence of the geometric variation of the rectangular stiffeners. Piscopo [8] studied analytically and numerically the uniaxial and biaxial elastic buckling of plates with no stiffeners. The analytical analysis used Shimpi's theory instead of the energies method, determining the corrected theoretical Euler buckling stresses. Its results were compared with the data obtained in the numerical analysis via FEM performed using the ANSYS® software. It was found that Shimpi's theory always provides, in relation to the energy method, results closer to the values obtained by the FEM analysis. Paik and Seo [9] performed the non-linear numerical analysis via FEM, to evaluate the ultimate stress of plates reinforced with longitudinal and transverse stiffeners subjected to biaxial compression, and also for combined biaxial compression of lateral pressure. It was possible to define which lateral pressure actions reduce the ultimate stress of the plate under predominantly compressive actions.

Therefore, the present work seeks to verify computational models for the application of FEM, using the ANSYS® software, with the purpose of evaluating the behavior of plates biaxially requested by compressive loads. The first numerical model will be verified based on the work presented in [8], where a thin plate without stiffeners is subjected to biaxial elastic buckling. The second model will be verified from the case presented by [9], where a thin plate with stiffeners is subjected to elasto-plastic buckling under biaxial compression.

2. Material and methods

For the developed computational models, this work used the ANSYS® software, based on FEM, which can be used in the most diverse types of engineering problems. For a structural analysis in ANSYS®, displacements and rotations (nodal degrees of freedom) are calculated numerically, and deformations and stresses are determined from these values [10].

For the development of the computational models of the present work, the finite element SHELL281 was adopted, being suitable for analyzing thin to moderately thick shell structures. The element has eight nodes with six degrees of freedom in each node: translations on the x, y and z axes and rotations on the x, y and z axes, being its formulation based on the first-order shear-strain theory (Reissner-Mindlin Theory) [11].

Based on El-Sawy and Nazmy [12], Helbig et al. [13], and Lima et al. [7] the computational models for biaxial elasto-plastic buckling of plates with stiffeners proposed in the present work was developed. For that, it is necessary to assume an imperfect initial configuration for the plate, which is obtained from its first elastic buckling mode. Therefore, a computational model for biaxial elastic plate buckling was preliminarily developed. For the numerical simulation of biaxial elastic buckling, the eigenvalue analysis was adopted. The finite element equilibrium equations for this type of analysis involve the solution of homogeneous algebraic equations, in which the lowest eigenvalue corresponds to the critical buckling load and the associated eigenvector represents the first buckling mode [14]. The formulation used includes both linear and non-linear terms, with the total stiffness matrix $[K]$ obtained by adding the conventional stiffness matrix for small deformations, $[K_E]$, with the geometric stiffness matrix $[K_G]$. The latter, $[K_G]$, does not depend only on the geometry, since the internal load $\{P_0\}$, existing at the beginning of the loading must be considered. Thus, the total plate stiffness matrix for a load level $\{P_0\}$, is given by [15]:

$$[K] = [K_E] + [K_G]. \quad (1)$$

In case the load reaches a level of $\{P\} = \lambda\{P_0\}$, where λ is a scalar, the stiffness matrix is written as:

$$[K] = [K_E] + \lambda[K_G]. \quad (2)$$

Then, the equilibrium equations, governing for the plate, can be written as:

$$\big[[K_E] + \lambda[K_G]\big]\{U\} = \lambda\{P_0\}, \quad (3)$$

where $\{U\}$ is the total displacement vector, which can be determined by:

$$\{U\} = \big[[K_E] + \lambda[K_G]\big] - \lambda\{P_0\}. \quad (4)$$

When λ reaches a critical value, the plate is subjected to a large increase in displacements without the load increasing. By mathematical definition it is possible to obtain the inverse matrix as the adjunct matrix divided by the coefficient determinant. So the displacements $\{U\}$ tend to infinity when:

$$det\,[[K_E] + \lambda[K_G]] = 0, \quad (5)$$

being *det* the determinant of the matrix.

The Eq. 5 represents an eigenvalue problem which, in which the smallest eigenvalue λ_1, corresponds to the critical buckling load given by:

$$\{P_{Cr}\} = \lambda_1\{P_0\}, \quad (6)$$

where P_{Cr} is the limit load that defines the biaxial elastic buckling of the plate, while the associated displacement vector $\{U\}$ defines the first buckling mode. Finally, it is worth noting that the eigenvalue problem was solved in ANSYS® using the Lanczos numerical method [16].

Then, for the numerical simulation of the biaxial elastoplastic buckling of plates, an elastic-perfectly plastic behavior (with no strain hardening) was assumed. In addition, as previously mentioned, a plate with an imperfect initial geometric configuration needs to be considered. For this, the indication made by El-Sawy and Nazmy [12] was adopted, which, from the first elastic buckling mode, defines the w_0 displacement as the maximum value of the initial imperfection, given by:

$$w_0 = b/2000, \quad (7)$$

where b iss the width of the plate.

Then, the ultimate biaxial buckling load acting on the plate is determined by gradually applying a reference load on the four sides of the plate, defined as:

$$P_y = \sigma_y t, \quad (8)$$

where σ_y is the yield stress of the material. For each load increase, the Newton-Raphson method is applied to determine the displacements that correspond to the equilibrium configuration of the plate. At the beginning of the loading step $i+1$, there is an out-of-balance load vector $\{\psi\}$, equal to the load increment $\{\Delta P\}$, between the vector of external loads, $\{P\}_{i+1}$, and the vector of nonlinear internal forces $\{F_{NL}\}$, that is equal to the previous external load vector $\{P\}_i$, then:

$$\{\psi\} = \{\Delta P\} = \{P\}_{i+1} - \{F_{NL}\} = \{P\}_{i+1} - \{P\}_i. \quad (9)$$

Therefore, the Newton-Raphson method reduce iteratively the out-of-balance load vector $\{\psi\}$ bellow a prescribed tolerance, by using the next equations:

$$\{\psi\}_{r+1} = \{P\}_{i+1} - \{F_{NL}\}_r, \quad (10)$$

$$[K_t]_r\{\Delta U\}_{r+1} = \{\psi\}_{r+1}, \quad (11)$$

$$\{U\}_{r+1} = \{U\}_r + \{\Delta U\}_{r+1}, \quad (12)$$

where $\{\psi\}_{r+1}$ is the updated out-of-balance load vector; $\{F_{NL}\}_r$ is the nonlinear internal forces vector at the iteration r; $[K_t]_r$ is the tangent stiffness matrix calculated as a function of the displacement vector $\{U\}_r$; $\{\Delta U\}_{r+1}$ is the updated displacement increment vector; and $\{U\}_{r+1}$ is the updated displacement vector.

The ultimate load of the plate is obtained if, in a specific loading step, convergence cannot be achieved in the iterative process. In this case, a finite increment in the displacement vector cannot be defined so that the out-of-balance load vector is nullified. This happens because no matter how large the strains and displacements, the internal dimensions and stresses cannot increase, as it requires necessary to balance the external loads, which indicates that the material of the plate reached its final strength [13,14,17].

From this, firstly it was verified the biaxial elastic buckling model, through a thin plate without stiffeners, like the one presented in Piscopo [8]. It is a simply-supported square plate, with length *a* and width *b* of 1000 mm, and thickness *t* of 10 mm, subjected to a biaxial compressive loading with the same intensity in both axes. The properties of the material used are: Young's modulus of 206 GPa and Poisson's ratio of 0.3. Fig. 1 shows the schematic representation of this plate.

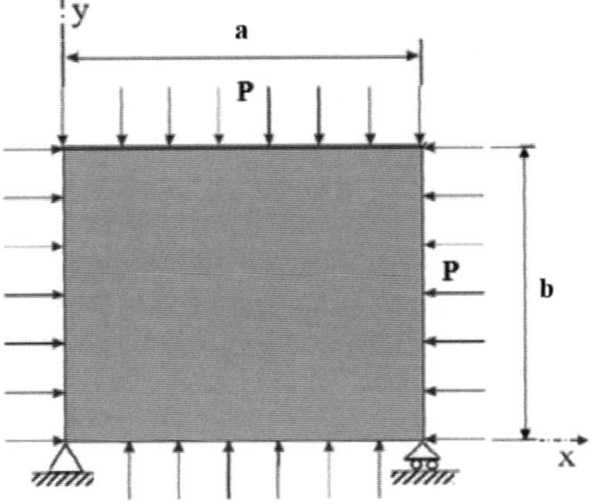

Figure 1. Plate simply supported on the edges, subjected to biaxial compression.

Then, for the verification of the computational model of biaxial elasto-plastic buckling, a case analyzed by Paik and Seo [9] was used. A biaxial compression analysis of a simply-supported rectangular plate with T-shaped stiffeners was performed. with the following dimensions: a = 4300 mm, b = 16300 mm, b_1 = 815 mm, t = 17.8 mm, h_w = 463 mm, t_w = 8 mm, b_f = 172 mm and t_f = 17 mm, with compressive biaxial loading of different intensities in each two axes ($\sigma_x : \sigma_y = 0.4 : 0.6$) [9]. The properties of the material used are: Young's modulus of 205.8 GPa, yield stress of 315 MPa and Poisson's ratio of 0.3. Fig. 2 shows the schematic representation of the plate and its stiffeners used in this verification.

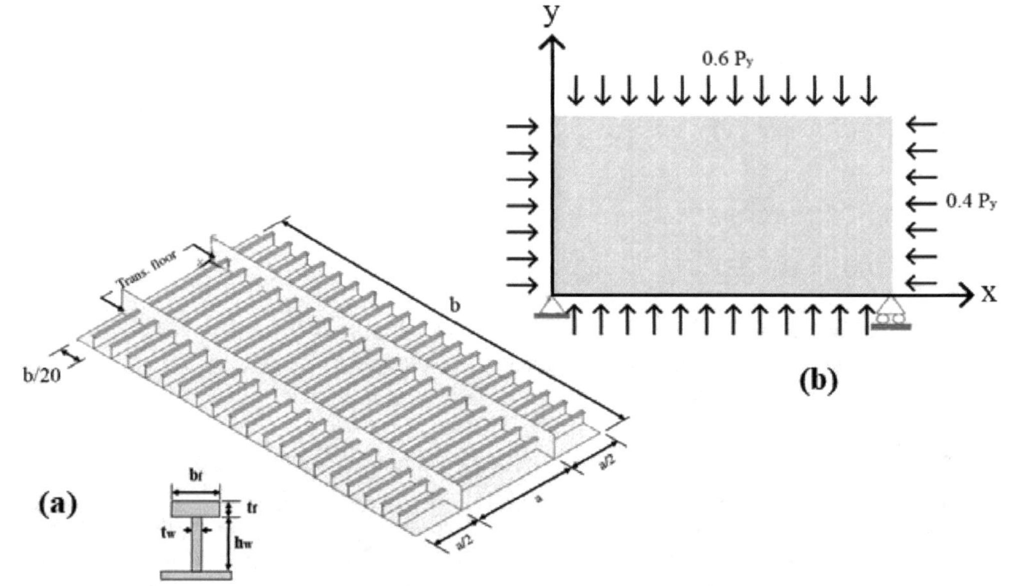

Figure 2. Simply-supported stiffened plate subjected to biaxial compression: (a) geometric configuration (Adapted from [9]), (b)computational domain.

3. Results and discussions

As described in item 2, the verification of the numerical model for biaxial elastic buckling was developed, since its result is used to determine the initial imperfect geometric configuration for the biaxial elasto-plastic buckling numerical model. In sequence, the verification of the computational model of biaxial elastoplastic buckling of a plate with stiffeners was performed.

3.1. Verification of the computational model for biaxial elastic plate buckling.

The case presented in Fig. 1 was analyzed by Piscopo [8], who obtained analytically a critical buckling load of P_{Cr} = 372 kN/m; and numerically a value of P_{Cr} = 370 kN/m. Thus, through the computational model proposed in this work, it was possible to numerically simulate the case of Fig. 1. For this, a mesh convergence test, varying the size of finite element from 100 to 10 mm. Therefore, it was obtained the variation of the critical load as the increase of mesh refinement, as can be seen in Fig. 3.

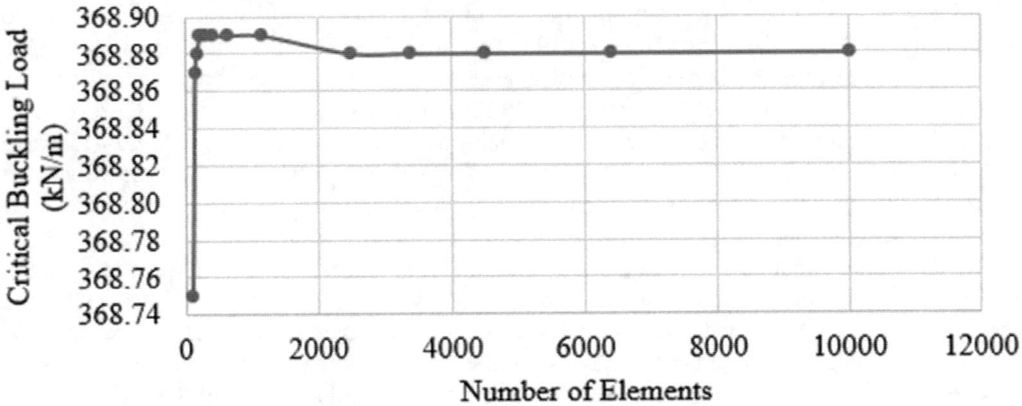

Figure 3. Mesh convergence test for the biaxial elastic buckling computational model for plates.

Inspecting the obtained results, it can be seen that from the mesh with 2500 elements, the numerical solution does not vary with the increase in the number of elements in the mesh, maintaining the critical load value of P_{Cr} = 368.88 kN/m. Therefore, the result obtained for the proposed model presents a difference of 0.84% and 0.30%, in relation to the results obtained by Piscopo [8] in his analytical and numerical models, respectively.

3.2. Verification of the computational model for biaxial elastoplastic buckling of plates with stiffeners. In Paik and Seo [9], see Fig. 2, it was obtained numerically the ultimate stress in x direction of 72.10 MPa, and the ultimate stress in y direction of 105.34 MPa. Then, by means the computational model developed in this work, the mesh convergence test was carried out, varying the size of each elements from 1000 to 100 mm, obtaining the variation of the ultimate buckling load, as can be seen be seen in Fig. 4.

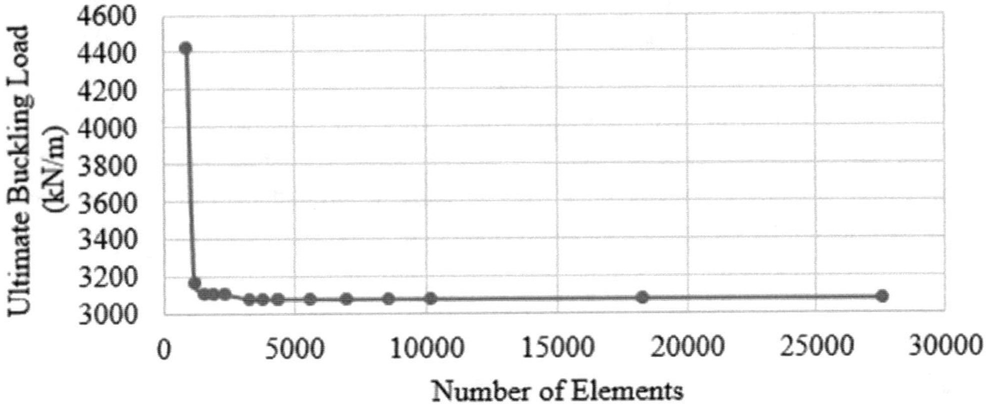

Figure 4. Mesh convergence test for the biaxial elasto-plastic buckling computational model for stiffened plates

One can note in Fig. 4 that from the mesh with 3800 elements the numerical solution does not vary with the increase in the number of elements in the mesh, maintaining the value for the ultimate load of P_U = 3083.80 kN/m. Calculating the ultimate stresses in x and y directions ($\sigma_{Ux} = 0.4P_U/t$ and $\sigma_{Uy} = 0.6P_U/t$), the ultimate stresses of σ_{Ux} =69.30 MPa and σ_{Uy} =103.95 MPa were, respectively, defined. Comparing the results obtained in this work with those obtained by Paik and Seo [9], it was achieved a difference of 3.89% in relation to the x direction, and a difference of 1.32% in relation to the y direction.

Conclusions

IOP Conf. Series: Materials Science and Engineering 1150 (2021) 012023 doi:10.1088/1757-899X/1150/1/012023

The present work proposed computational models for biaxial elastic and elasto-plastic plate buckling, developed using the Finite Element Method (FEM) in ANSYS® software. The biaxial elastic buckling model is used to determine the imperfect initial geometric configuration of the plate, which is used as an initial condition in the biaxial elasto-plastic buckling model.

Then, initially, the verification of the computational model of biaxial elastic buckling was performed, numerically simulating the problem proposed in Piscopo [8]. The numerical result for the critical load was compared with the analytical and numerical solutions of Piscopo [8], being identified a difference of 0.84% and 0.30%, respectively.

Afterwards, the verification of the computational model proposed for biaxial elastoplastic buckling was performed, comparing its results to those presented by Paik and Seo [9]. A plate with T-shaped transverse stiffeners and I-shaped longitudinal stiffeners was considered. A difference of 3.89% and 1.32% was found for the ultimate buckling stress in the x and y directions, respectively, comparing the numerical results obtained in the present study with those presented in Paik and Seo [9].

Therefore, it can be concluded, from the small differences found, that both proposed computational models were properly verified. In future work, we intend to use these models to carry out a study of geometric optimization of plates with stiffeners subjected to biaxial buckling.

Acknowledgments

G.R. Baumgardt thanks to the *Coordenação de Aperfeiçoamento de Pessoal de Nível Superior - Brasil* (CAPES) - Finance Code 001; and L.A.O. Rocha, E.D. dos Santos, and L.A. Isoldi thank to the *Conselho Nacional de Desenvolvimento Científico e Tecnológico - Brasil* (CNPq) - Processes: 307791/2019-0, 306024/2017-9, and 306012/2017-0, respectively.

References

[1] V. Birman, Solid Mechanics and Its Applications: Plate Structures, Springer, Netherlands, 2011.

[2] J.P.T.P. De Queiroz, M.L. Cunha, A. Pavlovic, L.A.O. Rocha, E.D. Dos Santos, G.S. Troina, L.A. Isoldi, Geometric Evaluation of Stiffened Steel Plates Subjected to Transverse Loading for Naval and Offshore Applications, J. Mar. Sci. Eng. 7 (2019) 7.

[3] M.L. Gambhir, Stability Analysis and Design of Structures, Springer-Verlag, Berlin, 2004.

[4] K.K. Kapur, B.J. Hartz, Stability of plates using the finite element method, Journal of the Engineering Mechanics Division. 92 (1966) 177-196.

[5] M.C. Xu, C. G. Soares, Experimental study on the collapse strength of wide stiffened panels, Mar. Struct, 30 (2013) 33–62.

[6] M.C. Xu, C. G. Soares, Comparison of calculations with experiments on ultimate strength of wide stiffened panels, Mar. Struct. 31 (2013) 82–101.

[7] J.P.S. Lima, M.L. Cunha, E.D. Dos Santos, L.A.O. Rocha, M.V. Real, L.A. Isoldi, Constructal Design for the ultimate buckling stress improvement of stiffened plates submitted to uniaxial compressive load, Engineering Structures. 203 (2020) 109883.

[8] V. Piscopo, Refined Buckling Analysis of Rectangular Plates Under Uniaxial and Biaxial Compression. World Academy of Science, Engineering and Technology International Journal of Mechanical and Mechatronics Engineering. 4 (2010) 1018-1025.

[9] J.K. Paik, J.K. Seo, Nonlinear finite element method models for ultimate strength analysis of steel stiffened-plate structures under combined biaxial compression and lateral pressure actions—Part II: Stiffened panels, Thin-Walled Structures. 47 (2009) 998–1007.

[10] Ansys. Analysis Guide. Cannonsburg: Ansys Inc, 2004.

[11] ANSYS Help Viewer. Release 18.1 - © ANSYS, Inc. All rights reserved.

[12] K.M. El-Sawy, A.S. Nazmy, M.I. Martini, Elasto-Plastic Buckling of Perforated Plates under Uniaxial Compression. Thin-Walled Structures. 42 (2004) 1083-1101.

[13] D. Helbig, C.C.C. Da Silva, M.V. Real, E.D. Dos Santos, L.A. Isoldi, L.A.O. Rocha, Study About Buckling Phenomenon in Perforated Thin Steel Plates Employing Computacional Modeling and Constructal Design Method, Latin American Journal of Solids and Structures. 13 (2016) 1912-1936.

[14] E. Madenci, I. Guven, The Finite Element Method and Applications in Engineering Using ANSYS®, second ed., Boston: Springer, 2015.
[15] J.S. Przemieniecki, Theory of Matrix Structural Analysis, Dover Publications, 1985.
[16] Ansys. User's Manual (version 10.0), Swanson Analysis System Inc, Houston, 2005.
[17] J.P.S. Lima, L.A.O. Rocha, E.D. Dos Santos, M.V. Real, L.A. Isoldi, Constructal design and numerical modeling applied to stiffened steel plates submitted to elasto-plastic buckling, Proceedings of the Romanian Academy - Series A: Mathematics, Physics, Technical Sciences, Information Science. 19(Special Issue) (2018) 195-200.

IOP Conf. Series: Materials Science and Engineering 1150 (2021) 012024 doi:10.1088/1757-899X/1150/1/012024

Investigation of the Effects of the Classification of Building Stock Geometries Determined Using Clustering Techniques on the Vulnerability of Galvanized Iron Roof Covers Against Severe Wind Loading

Tan, Liezl Raissa E., Acosta, Timothy John S.*, Gumaro, Joshua Joseph C., Agar, Joshua C., Tingatinga, Eric Augustus J., Plamenco, Dean Ashton D., Ereno, Mary Nathalie C., Musico, John Kenneth B., Pacer, Jihan S., Baniqued, Julius Rey D., Hernandez, Jaime Jr. Y., Villalba, Imee Bren O.

[1]Insitute of Civil Engineering, University of the Philippines Diliman, Philippines

E-mail: tsacosta@up.edu.ph

Abstract. In the risk assessment of buildings against severe wind loading, the vulnerability component of risk is highly affected by the geometry of the structure. Parameters such as the height, width, length, aspect ratio, eaves length, and roof slope, affect the pressure distribution around the structure, which in turn affects the response of galvanized iron (GI) roof covers to wind loadings. In developing countries, there is a large variation in the building geometric parameters which poses a challenge in determining the archetypes that would best represent the building population for risk assessment. This paper aims to develop and propose a method in determining the building archetypes based on its geometry. The hierarchy for grouping of geometries started with the roof type. These were gable type roofs, mono-slope type roofs and hip type roofs. The building datasets per roofing type were then clustered using a two-stage approach involving Hierarchical and K-means clustering which were based on the aforementioned geometric parameters. These algorithms will aggregate buildings having similar sets of geometric parameters but the number of clusters must be specified. In order to determine the optimal number of clusters, this study employs various validation tools or measures namely – dendrograms, variation of the variance ratio criterion (VRC) across number of clusters, validity indices such as, Davies Bouldin, Silhouette and Calinski Harabasz, and the elbow method. Although guided by these validation methods, the final selection of the number of clusters were determined considering computational time and resources. To define an archetype, the mean values of each parameter per cluster were selected. Resulting to 5, 3, and 3, archetypes for gable, hip, and mono-slope roof buildings, respectively. The selection of the archetype was further evaluated by investigating its effects on the vulnerability of GI roof covers in order to see how distinct each archetype would behave. A Kruskal Wallis test on the vulnerability curves of the different building archetypes showed that there is a significant difference between vulnerability curves under a roof shape category, which reinforces the distinction between the selected building archetypes.

1. Introduction

The determination of the building stock is a key parameter in the field of risk analysis. Albayrak et. al (2015) were able to classify buildings into existing building stock for rapid seismic risk assessment [1]. The classification of building stock helps most especially in the classification of local housing construction wherein these are non-engineered buildings. Parameters such as the main structural frame, load-bearing systems, connections between structural elements among others, are used to identify basic typologies for certain countries [2]. One classification system was developed for the Prompt Assessment of Global Earthquakes for Response (PAGER) system. This system is used to estimate an earthquake's impact in order to guide governments, insurance agencies and relief organizations in post-disaster relief operations and strategies [3]. This system made use of a global building inventory that was developed through the use of different inventory data sources worldwide [4]. Databases such as the UN statistical database on global housing [4], the United Nations Human Settlement Program (UN-HABITAT) database [4], World Housing Encyclopedia (WHE) database [5] have information on the housing dwelling type, construction materials for wall, roof or floors, structural systems and other parameters that contribute to the vulnerability of building to seismic activity [4]. Aside from these databases, FEMA was able to develop HAZUS standard model building classification schemes. These building stock classifications were not only used for seismic risk assessment but were used for multi-hazard loss estimations.

One of the hazards which HAZUS assesses is severe wind events. For their methodology, two building stock classification methods were used. The first method was through the inspection of aerial photography samples of Dade, Broward and Palm Beach of Southeast Florida. It was concluded that, for commercial building stocks, simple rectangular-shaped building models were adequate for damage simulation which came from random sampling from the aerial photographs wherein they observed that 80% of the samples accounted for rectangular or L shaped plan areas. For the residential building stock, they used three simple default classifications which were the single-story gable, single story hip and two-story gable. It was also noted that no attempt was made to quantify the plan shapes. The other method made use of contractor surveys to determine the distribution of GI roof cover types. The method was able to come up with 13 building stock classifications [6]. Each building stock classification had multiple combinations of building envelope components that contributed to the vulnerability against wind loadings. For the building geometries used to determine the wind loads were based on expert opinion for each building stock classification [6,7]. The different building geometries which were derived using expert opinion may also affect the determination of the wind loads which in turn will affect the resulting vulnerability of the building stock.

With the advancement of computational fluid dynamics (CFD), many researches have been done to explore the effects of geometric parameters on the wind loads of a building [8,13,14,15]. In the application of risk assessment, the use of CFD to individually analyze the configuration of the population of buildings is very complicated and computationally heavy. Thus, there is a need to explore the use of a building geometry that would statistically best represent the population of buildings being assessed and will also improve the risk assessment of the population of buildings being considered. One method is to use clustering techniques on diverse data to come up with a representative ensemble for the population of buildings. Since the wind pressures on the building will be computed using computational fluid dynamics, the objective of determining the building geometry archetypes, instead of individual buildings of the population, will significantly reduce the computational cost.

The use of clustering techniques in the classification of building stock for different applications has been prominent in the past decade. The technique has been applied to cluster buildings; based on their sensitivity to retrofitting measures [16], in the context of air conditioning electricity demand flexibility

[17], performance in energy savings [18], energy efficiency and emission reduction [19], and many other areas. With this in mind, this paper seeks to investigate the use of clustering techniques to develop building stock geometries for risk assessment of low-rise buildings against severe wind loadings.

This proposed methodology was implemented in a pilot study to assess the risk of the province of Cebu, Philippines against severe wind loadings. The pilot study is part of the Philippine government's national effort to produce scientifically based vulnerability profiles which would help in disaster risk and reduction strategies. The greater metro manila area risk analysis project (GMMA-RAP) [20] is a precedent of the current pilot study wherein key government agencies and institutions such as the Philippine Atmospheric, Geophysical and Astronomical Services Administration (PAG-ASA), Philippine Institute of Volcanology and Seismology (PHIVOLCS) and Institute of Civil Engineering, University of the Philippines Diliman (UPD-ICE) were involved. And given a lack of a high-resolution exposure database, this study was performed to improve the risk analysis methodology.

2. Methodology

2.1. Parameter Identification.

Various researches have shown that the roof shape, slope, length, width, height [9,10,11,12] have significant effect on the pressures for low-rise structures. The hierarchy of the parameters was patterned after the aerodynamic database developed by Tokyo Polytechnic University (TPU) [21]. The dataset was then classified by the roof shapes.

2.2. Clustering Analysis.

Clustering techniques group data points into clusters in a way that the data points in the same cluster are similar to one another compared to data points in another cluster [22]. This paper employed a two-stage approach involving Hierarchical and K-means clustering algorithms employed in MatLAB. The two algorithms complemented each other in a way that k-means clustering, known for its efficiency in large datasets, works best when a non-random starting point is specified, which is provided by the cluster centroids obtained from the hierarchical clustering. A series of the number of clusters were run using the k-means approach where each building data point was assigned to the nearest cluster. The cluster centroids were recomputed and data points reassigned to minimize the variability within clusters whilst maximizing the variability between clusters. This approach is similar to the ones employed in the study on electricity usage for consumer archetypes [22].

2.3. Evaluation of the results.

In order to assess the optimum number of clusters, various validation methods were utilized. Visual methods such as the use of dendrograms and the elbow method, were used to roughly assess the distinction across clusters. The elbow method plots the within-cluster-sum-of-squared error (WSS), for every number of clusters considered. The number of clusters corresponding to the elbow of the graph is regarded as the optimum number of clusters. WSS is computed as follows,

$$WSS = \sum_{k}^{C} \sum_{i \in k} (x_i - \mu_k)^2 \quad (1)$$

Where x_i is an element within the kth cluster C_k with mean value equal to μ_k. Thus WSS takes the summation of the summation of the sum-of-square error (SSE) within the cluster which is then summed across all clusters C.

Internal validity indices were also used to evaluate the clustering performance. These are the Silhouette index, Davies Bouldin Index and the Calinski Harabasz Index. The silhouette width indicates the belongingness of a datapoint in a cluster it was assigned. Values ranges from -1 to 1 wherein a value close to 1 means that the "within" dissimilarity is much smaller than the "between" dissimilarity, hence it is said to be well-clustered. Davies Bouldin index is a ratio of the total scatter within the cluster to the separation between clusters, and hence a smaller value is regarded as good clustering. Calinski Harabasz is the ratio of the overall between-cluster variance to the overall within-cluster variance multiplied by the factor ratio between the difference of the total data points and the number of clusters to the number of clusters minus 1. There is no acceptable cut-off value, but the higher the index the better.

Another validation method makes use of the variance ratio criterion (VRC) [24,25]. F value were determined for each parameter, p, per number of clusters, C. The F value, F_p, measures the ratio of variability of each parameter, p, between clusters, k, given by the Between-Group-Sum-of-Square Error (BGSS) and the variability within a cluster or WGSS. [23].

$$F_p = \frac{\frac{BGSS}{C-1}}{\frac{WGSS}{n-C}} \tag{2}$$

$$BGSS = \sum_k n_k \|\mu_k - \mu\|^2 \tag{3}$$

$$WGSS = \sum_k \sum_{i \in k} \|x_i - \mu_k\|^2 \tag{4}$$

Where μ, represents the centroid of all the n number of datapoints,x_i, while μ_k represents the centroid within clusters of n_k datapoints.

The F value of each parameter can be used to identify which parameter greatly affects the distinction between clusters. So, for a number of clusters considered, the higher the F value of a parameter the higher its influence to the clustering process, and therefore can be regarded as a critical clustering parameter. In relation to determining the optimum number of clusters, the F values were totaled to get the VRC for every number of clusters. The higher the value of the VRC implies that either clusters are highly distinct from each other or that the data points within each cluster is highly similar with each other, and thus a high value of VRC is preferable in the selection of the number of clusters. However, the VRC value generally increases as the number of clusters approaches the total number of datapoints. To aid in selecting an efficient number of clusters, the VRC values were then compared to its neighboring number of clusters using the equation for ω_k, which quantifies the difference between the deviations of VRC from the current cluster to the succeeding and preceding clusters.

$$\omega_k = (VRC_{k+1} - VRC_k) - (VRC_k - VRC_{k-1}) \tag{5}$$

A positive value for the first term indicates that the current cluster performs worse than the succeeding cluster while a positive value for the second term indicates that the current cluster performs better than the previous cluster. Therefore, the more negative the value of the ω_k indicates a more efficient selection of the number of clusters.

2.4. Roofing Vulnerability Analysis.

In order to illustrate the use of the above geometries, the vulnerability curves for the roof sheeting were determined in this paper. The framework was patterned after the GMMA-RAP simulation methodology [20]. The archetypes developed in the clustering analysis served as the representative geometries that would be subjected to computational fluid dynamics. The damage ratio was finally derived by determining the percent area of the roofing sheets that exceeded the threshold capacity and dividing this by the total area. The vulnerability curve for each building archetype was then derived by getting the mean of the models accounting for different wind directions and plotting by the corresponding wind speed.

3. Results: Case Study of Cebu, Philippines

3.1. Clustering Analysis.

For the pilot study area, which is Cebu, Province, the parameters of interest were surveyed by a team coming from the Institute of Civil Engineering, University of the Philippines Diliman. The target areas in Cebu were chosen based on the analysis of the construction material typology per municipality wherein the data on the construction materials came from the housing tables of the Philippine Statistic Authority (PSA) [26]. A total of 179 buildings from 15 municipalities of the province were gathered. The prominent roof shapes were gable roofs, mono-slope roofs, and hip roofs which comprised 50.8%, 24.58%, and 12.89% of the data, respectively. Hence these were selected as the three main groups used for the cluster analysis.

The figure below shows the dendrograms produced by the hierarchical clustering algorithm. The centroids of these results were used as the input cluster centroids for the k-means clustering algorithm. Each number in the horizontal axis represents the building ID number and hence represents one building data point. While the vertical axis represents the mean values of the parameters or also termed as cluster centroid. The height of the links in a dendrogram is indicative of the distinction between clusters. Each color represents a cluster in a group of a particular roof shape, and consequently shows how big of the building population a cluster represents.

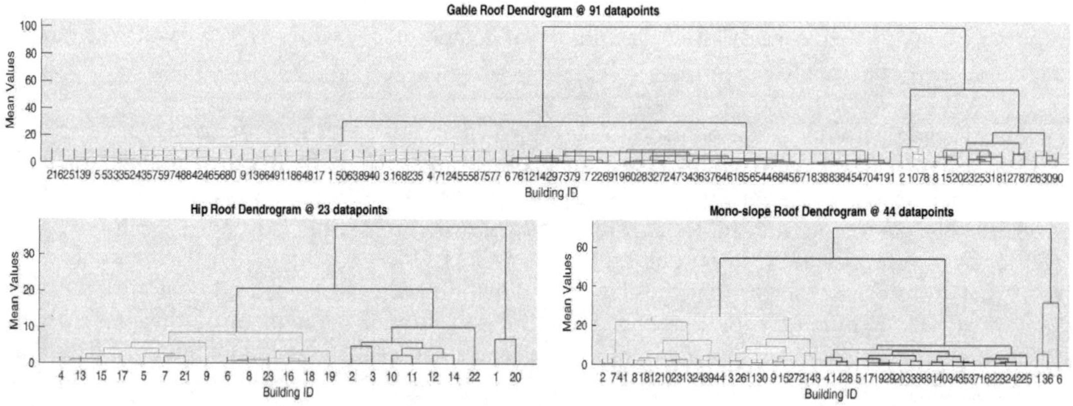

Figure. 1 Dendrograms from Hierarchical Clustering of different roof shapes using Ward's Method.

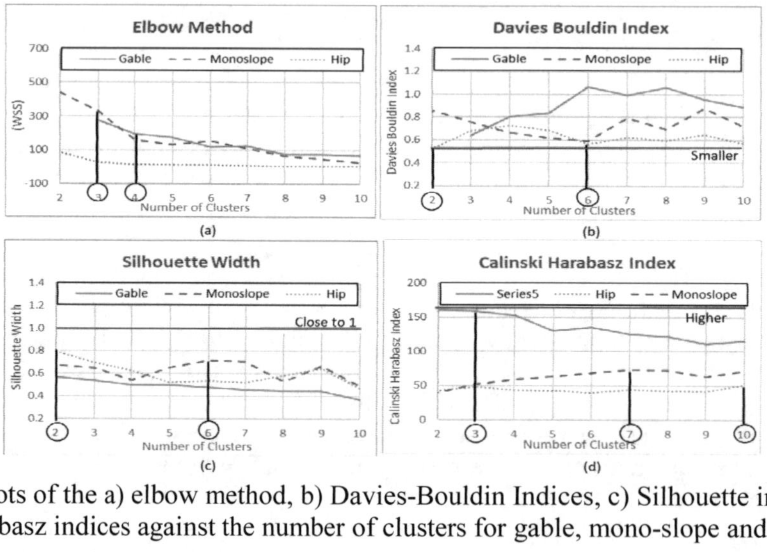

Figure. 2 Plots of the a) elbow method, b) Davies-Bouldin Indices, c) Silhouette indices, and d) Calinksi-Harabasz indices against the number of clusters for gable, mono-slope and hip roof types.

IOP Conf. Series: Materials Science and Engineering 1150 (2021) 012024 doi:10.1088/1757-899X/1150/1/012024

The validation indices from various methods provided the team a range of number of clusters to select from and also gives a measure of how distinct these clusters are from each other yet contains data points that are similar within each cluster. Consideration of the computational time and resources were also a major influence in the selection of the final number of clusters.

The ω_k values were obtained for 3 to 10 clusters for all roof shapes and are shown in the following figure.

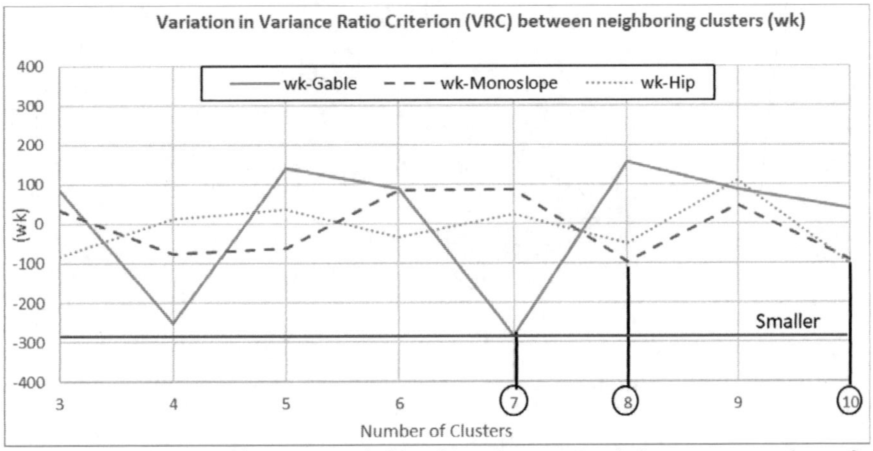

Figure. 3 Variation in VRC between neighboring clusters (ω_k) for every number of clusters.

A selection between 2 to 7, 4 to 8 and 3 to 10 number of clusters were considered for gable, mono-slope and hip roof types, respectively. Considering the roof type composition in the building and computational time and resources the author's settled at 5, 3 and 3 number of clusters for gable, mono-slope and hip roof types, respectively. In order to understand the characteristics of the population of buildings, the relative contribution of each clustering parameter to the development of the clusters was investigated. The F values of each parameter were ranked to see the importance of each parameter to the clustering algorithm.

Table 1. Rank, *F*, *VRC* and ω_k values for parameters of the cluster solutions of different roof shapes

Parameter	Gable Clusters *k* =5		Hip Clusters *k* = 3		Mono-slope Clusters *k* = 3	
	Rank	F	Rank	F		F
Height	3	12.79	3	6.66	3	14.32
Length	1	263.88	1	108.93	1	71.38
Width	2	142.69	2	50.65	2	45.51
Slope	6	0.77	6	2.84	4	4.44
Eaves-L	5	1.16	5	4.36	5	1.48
Eaves-S	4	5.91	4	4.76	6	1.31
VRC	-	427.22	-	178.19	-	138.43
wk	-	139.094	-	-83.80	-	-276.87

The table above shows us that the primary parameters that affected the formation of clusters are the length (ranked 1st) and width (ranked 2nd) of the building geometries. This implies that the building population can be categorized more by their plan area rather than their roof shape parameters. This also tells us that the construction practice for the roof slope, eaves and length has a low variability in the pilot study area.

The final geometry clusters produced through the clustering techniques are now considered as the archetypes which will be used to determine the wind loads on the roofing sheets. The building archetypes for the different roof shapes can be found in table 2 to 4. The illustration in figure 4 gives a reference on how the different parameters are measured.

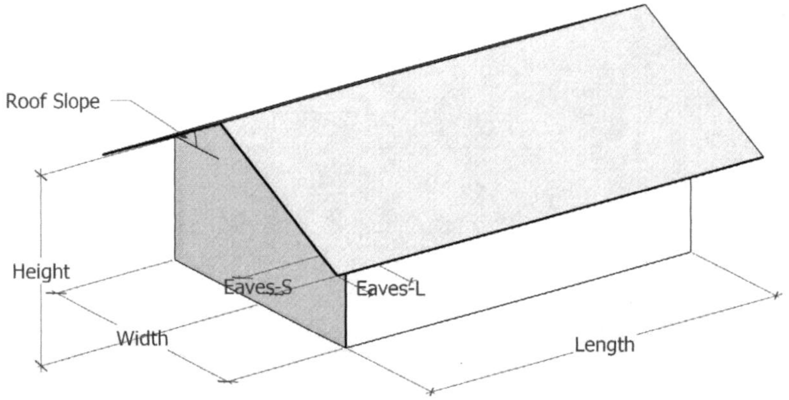

Figure. 4 Illustration of parameter measurements for gable roof archetypes

Table 2. Centroids for Gable K-means clusters

ARCHETYPE LABEL	HEIGHT	LENGTH	WIDTH	ROOF SLOPE	EAVES-L	EAVES-S
G1	4.08	10.50	5.07	15.33	1.01	0.50
G2	8.50	41.67	27.00	19.00	1.10	1.90
G3	4.26	5.97	4.62	19.00	0.85	0.58
G4	8.75	22.00	16.35	19.00	1.22	0.25
G5	5.84	11.6	8.76	19.5	0.68	0.58

Table 3. Centroids for Hip K-means clusters

ARCHETYPE LABEL	HEIGHT	LENGTH	WIDTH	ROOF SLOPE	EAVES-L	EAVES-S
H1	3.95	7.37	5.70	24.20	0.83	0.70
H2	8.90	24.95	17.45	38.31	1.45	1.45
H3	5.90	13.42	8.75	21.03	1.03	0.875

Table 4. Centroids for Mono-slope K-means clusters

ARCHETYPE LABEL	HEIGHT	LENGTH	WIDTH	ROOF SLOPE	EAVES-L	EAVES-S
M1	5.42	6.92	4.78	8.83	0.70	0.20
M2	9.00	18.23	11.00	2.50	0.50	0.10
M3	20.00	50.00	30.00	0.00	0.00	0.00

3.2. Vulnerability Results.

The simulations were done for wind speeds from 10 m/s to 100 m/s and considered different orientations of the wind relative to the building surfaces. Due to the symmetry of the archetypes, only angles between 0 degrees to 90 degrees at increments of 15 degrees were considered. The various combinations gave a total of 770 models that were simulated with 350 models for the 5 gable roof archetypes, 210 models for the 3 hip roof archetypes and 210 for the 3 mono-slope roof archetypes. All CFD simulations were implemented through the software ANSYS CFX. The GI roof cover failure which was primarily characterized by pullout strength of 4.3 kPa [27] was compared to the positive and negative wind pressures developed on the building archetypes. The simulated data points were then fitted into a lognormal cumulative distribution function (CDF) using a nonlinear regression analysis that minimized the sum of squared estimate of errors (SSE) to represent the final vulnerability curves. The final vulnerability curves are then shown in figure 5.

Table 5. Lognormal cumulative distribution function parameters

CDF Parameters	BUILDING ARCHETYPES										
	G1	G2	G3	G4	G5	H1	H2	H3	M1	M2	M3
Mean	4.92	4.86	4.87	4.67	4.69	4.92	4.89	4.83	4.65	4.54	4.38
Std	0.31	0.37	0.30	0.23	0.21	0.29	0.29	0.32	0.22	0.21	0.36

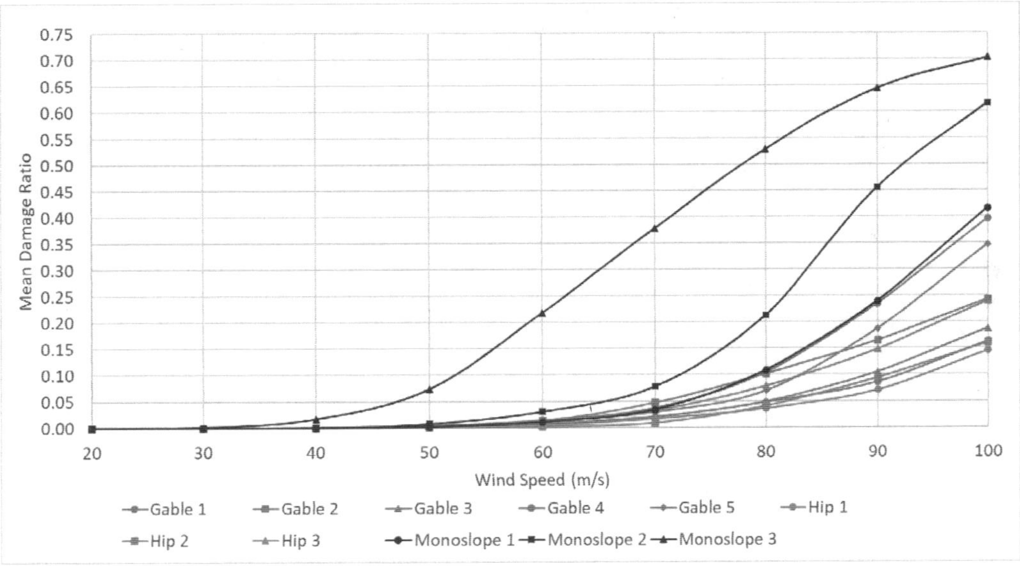

Figure. 5 Roof Sheeting Vulnerability for different building archetypes

A Kruskal-Wallis test was used to determine if the vulnerability curves for each building archetype were significantly different from one another. For gable, hip and mono-slope roof building archetypes, it was observed that for wind speeds from 60 m/s – 100 m/s, there was a significant difference ($p = .001$) for the vulnerability curves between the 5 building archetypes at a level of significance of 5%. For hip roof building archetypes, the same conclusion of a significant difference ($p = .001$) was also observed between the three building archetypes. And for the mono-slope building archetypes, the vulnerability curves were also significantly ($p < .0001$) different from each other. These results show that the use of building archetypes developed from the two-stage clustering algorithm has a significant effect on the vulnerability analysis of roof sheeting against severe wind loadings. The use of multiple building archetypes that best represent the population of buildings considers the variability of the geometric parameters of the area being assessed. The analysis can be extended to investigate the vulnerability curves that also consider other building envelope components aside from the roofing sheets that are also vulnerable to severe wind loadings.

4. Conclusion

This paper presents the application of clustering techniques in determining the building geometry archetypes which will be used for vulnerability and risk assessments. Specifically, the vulnerability assessment for galvanized iron roof covers against severe wind loading was evaluated. The paper implemented a two-stage clustering approach wherein Hierarchical and K-means clustering algorithms were used to develop clusters for the population of buildings in Cebu, Philippines. A total of 179 buildings were considered and subdivided into 3 main categories based on roof shape namely the gable, mono-slope and hip. Cluster analysis was employed on buildings in each roof shape category considering the parameters, length, width, height, roof slope, and eaves lengths. The analysis considers various numbers of clusters where cluster centroids and cluster assignments were obtained. The clusters were then evaluated using validation scores for cluster analysis namely the elbow method, silhouette index, Davies-Bouldin, Calinski Harabasz and the variation in VRC between neighboring clusters (ω_k). The validation scores were used to serve as a guide in the selection of the number of clusters to define the archetypes per roof type. Considering the composition of building population and the computational time and resources, the final number of clusters were selected for each roof shape. These are 5, 3, and 3 for gable, mono-slope and hip roof type, respectively. The cluster centroids for the selected number of clusters per roof type were then used as the dimensions for the building archetypes to be subjected to vulnerability analysis. For each building archetype, the simulated damage ratios for every wind speed data were fitted with lognormal cumulative distribution functions in obtaining the vulnerability curves. The vulnerability curves for the gable, hip and mono-slope building archetypes were observed to have a significant difference within each respective roof category at a level of significance of 5%. This shows that the consideration of the building archetypes from a sample population will significantly affect the results of the vulnerability analysis for wind speeds greater than 60 m/s. Further studies can be done to investigate the effect on vulnerability curves that also consider different building components that are vulnerable to severe wind loadings.

Acknowledgements

This work is part of the project entitled "Enhanced Vulnerability Curves for Different Building Types in the Philippines" under the program "Severe Wind Risk Assessment of Cebu City" wherein this project received funding from the Department of Science and Technology (DOST) through the Grants-In-Aid (GIA) program. The program was implemented in collaboration with the Philippine Atmospheric, Geophysical and Astronomical Services Administration (PAG-ASA) and Philippine Institute of Volcanology and Seismology (PHIVOLCS).

References

[1] Albayrak, Uğur, Mehmet Canbaz, and Gülçağ Albayrak. "A rapid seismic risk assessment method for existing building stock in urban areas." Procedia engineering 118 (2015): 1242-1249.

[2] Ngoma, Ignasio, Innocent Kafodya, Panos Kloukinas, Viviana Novelli, John Macdonald, and Katsuichiro Goda. "Building classification and seismic vulnerability of current housing construction in Malawi." Malawi Journal of Science and Technology 11, no. 1 (2019): 57-72.

[3] Earle, Paul S., David J. Wald, Kishor S. Jaiswal, Trevor I. Allen, Michael G. Hearne, Kristin D. Marano, Alicia J. Hotovec, and Jeremy M. Fee. "Prompt Assessment of Global Earthquakes for Response (PAGER): A system for rapidly determining the impact of earthquakes worldwide." US Geological Survey Open-File Report 1131 (2009): 15.

[4] Jaiswal, Kishor, and David Jay Wald. Creating a global building inventory for earthquake loss assessment and risk management. US Geological Survey open-file report-1160. Reston, VA: US Geological Survey, 2008.

[5] EERI, 2007, World Housing Encyclopedia (http://www.world-housing.net/)

[6] FEMA. "HAZUS-MH 2.1 hurricane model technical manual." (2012).

[7] Vickery, Peter J., Peter F. Skerlj, Jason Lin, Lawrence A. Twisdale Jr, Michael A. Young, and Francis M. Lavelle. "HAZUS-MH hurricane model methodology. II: Damage and loss estimation." Natural Hazards Review 7, no. 2 (2006): 94-103.

[8] Singh, Jagbir, and Amrit Kumar Roy. "Effects of roof slope and wind direction on wind pressure distribution on the roof of a square plan pyramidal low-rise building using CFD simulation." International Journal of Advanced Structural Engineering 11, no. 2 (2019): 231-254.

[9] Alrawashdeh, Hatem, and Ted Stathopoulos. "Wind pressures on large roofs of low buildings and wind codes and standards." Journal of Wind Engineering and Industrial Aerodynamics 147 (2015): 212-225.

[10] Kasperski, Michael. "Design wind loads for low-rise buildings: a critical review of wind load specifications for industrial buildings." Journal of wind engineering and industrial aerodynamics 61, no. 2-3 (1996): 169-179.

[11] Holmes, John D., Yukio Tamura, and Prem Krishna. "Wind loads on low, medium and high-rise buildings by Asia-Pacific codes." In The 4th International Conference on Advances in Wind and Structures, Jeju, Korea. 2008.

[12] Prasad, Deepak, Tuputa Uliate, and M. Rafiuddin Ahmed. "Wind loads on low-rise building models with different roof configurations." International Journal of Fluid Mechanics Research 36, no. 3 (2009).

[13] Janajreh, Isam. "CFD Analysis of Wind Loads on Permeable Low-Rise Structures." In Applied Mechanics and Materials, vol. 152, pp. 1814-1820. Trans Tech Publications Ltd, 2012.

[14] Tominaga, Yoshihide, Shin-ichi Akabayashi, Takuya Kitahara, and Yuki Arinami. "Air flow around isolated gable-roof buildings with different roof pitches: Wind tunnel experiments and CFD simulations." Building and Environment 84 (2015): 204-213.

[15] Atmaca, M. U. S. T. A. F. A. "Wind tunnel experiments and CFD simulations for gable-roof buildings with different roof slopes." Acta Physica Polonica A, vol 135 (2019): 690-3.

[16] Geyer, Philipp, Arno Schlüter, and Sasha Cisar. "Application of clustering for the development of retrofit strategies for large building stocks." Advanced Engineering Informatics 31 (2017): 32-47.

[17] Patteeuw, Dieter, Gregor P. Henze, Alessia Arteconi, Charles D. Corbin, and Lieve Helsen. "Clustering a building stock towards representative buildings in the context of air-conditioning electricity demand flexibility." Journal of Building Performance Simulation 12, no. 1 (2019): 56-67.

[18] Lee, Yunseok, Pierre Boisson, Mathieu Rivallain, and Olivier Baverel. "Application of clustering technique for the segmentation of building stocks to renovate." 2016.

[19] Geyer, Philipp, and Arno Schlueter. "Performance-based clustering for building stock management at regional level." In Proceedings of the International Conference 'Smart Energy Regions', pp. 230-241. Cardiff University, 2016.

[20] UP Diliman Institute of Civil Engineering, Development of Vulnerability Curves of Key Building Types in the Greater Metro Manila Area, Philippines (2014).

[21] TPU Aerodynamic database (http://www.wind.arch.t-kougei.ac.jp/system/eng/)

[22] Hughes, Mark., Moreno, Jorge Garcia. “Further Analysis of Data from the HEUS Consumer Archetypes.” Final report for the Department of Energy and Climate Change and Department for the Environment Food and Rural Affairs. (2013).

[23] Nikolaou, Triantafyllia G., Dionysia S. Kolokotsa, George S. Stavrakakis, and Ioannis D. Skias. "On the application of clustering techniques for office buildings' energy and thermal comfort classification." IEEE Transactions on Smart Grid 3, no. 4 (2012): 2196-2210.

[24] Caliński, Tadeusz, and Jerzy Harabasz. "A dendrite method for cluster analysis." Communications in Statistics-theory and Methods 3, no. 1 (1974): 1-27.

[25] Milligan, Glenn W., and Martha C. Cooper. "An examination of procedures for determining the number of clusters in a data set." Psychometrika 50, no. 2 (1985): 159-179.

[26] Philippine Statistics Authority. 2015 FIES Housing Tables (https://psa.gov.ph/content/2015-fies-housing-tables)

[27] Baskaran, Bas A., Steven Kee Ping Ko, and Suda Molleti. "A novel approach to estimate the wind uplift resistance of roofing systems." Building and environment 44, no. 4 (2009): 723-735.

AUTHOR INDEX